Reparative Environmental Justice in a World of Wounds

Reparative Environmental Justice in a World of Wounds

Ben Almassi

LEXINGTON BOOKS
Lanham • Boulder • New York • London

Published by Lexington Books
An imprint of The Rowman & Littlefield Publishing Group, Inc.
4501 Forbes Boulevard, Suite 200, Lanham, Maryland 20706
www.rowman.com

6 Tinworth Street, London SE11 5AL, United Kingdom

Copyright © 2021 The Rowman & Littlefield Publishing Group, Inc.

All rights reserved. No part of this book may be reproduced in any form or by any electronic or mechanical means, including information storage and retrieval systems, without written permission from the publisher, except by a reviewer who may quote passages in a review.

British Library Cataloguing in Publication Information Available

Library of Congress Cataloging-in-Publication Data

Names: Almassi, Ben, 1979- author.
Title: Reparative environmental justice in a world of wounds / Ben Almassi.
Description: Lanham : Lexington Books, [2020] | Includes bibliographical references and index. | Summary: "Reparative Environmental Justice in a World of Wounds questions how we can repair human and biotic relationships damaged by environmental injustice, climate change, animal exploitation, and ecological destruction by advertising the merits of a reparative approach to environmental justice and a critical assessment of the challenges that come with it"—Provided by publisher.
Identifiers: LCCN 2020039209 (print) | LCCN 2020039210 (ebook) |
  ISBN 9781498592062 (cloth) | ISBN 9781498592079 (epub)
  ISBN 9781498592086 (pbk)
Subjects: LCSH: Environmental justice. | Restoration ecology. | Epistemics.
Classification: LCC GE220 .A46 2020 (print) | LCC GE220 (ebook) |
  DDC 304.2/8—dc23
LC record available at https://lccn.loc.gov/2020039209
LC ebook record available at https://lccn.loc.gov/2020039210

Contents

Preface		vii
1	Justice after the Dam Breaks	1
2	Environmental Injustice and Its Amelioration	19
3	A Relational Revaluation of Ecological Restoration	37
4	Animal Ethics and Contexts of Interspecies Repair	53
5	Climate Change and Intergenerational Reparative Justice	69
6	Traditional Ecological Knowledge and Reparative Epistemic Justice	87
7	Reparative Environmental Justice in the Chicago Wilderness	109
8	Alone and Together in a World of Wounds	137
Bibliography		143
Index		171
About the Author		175

Preface

The past fifty years have seen incredible work in the fields of environmental ethics and political philosophy, both in ideal theory and in practical application. In this same period, we have also witnessed the birth of US environmental justice and animal-rights movements, and a growing recognition of the power of restorative justice. Yet, we in philosophy, have attended too rarely to the intersections of these rich fields and movements and to the distinctive contributions we can make there. Contemporary moral, political, and environmental philosophies have had much to say about justice, value, rights, obligation, and duty, and comparatively less to say on the role of philosophy *after* injustice or wrongdoing, about what to do given perpetrations of injustice and failures of duty and obligation. This relative silence is especially odd in environmental contexts, where ethical and social-political inadequacies are not the exception but the rule, whether we are considering environmental racism and health inequity, animal abuse and exploitation, ecosystem degradation, or climate change and intergenerational injustice. This is not to deny the value of ideal theory for, say, animal rights or intergenerational ethics, but rather to emphasize the need for environmental philosophy that speaks to what comes after ideal theory breaks down.

"Moral philosophers in the twentieth century have often liked to characterize ethics as answering the question, 'What ought I to do?' which implies a set of choices on a fresh page," writes Margaret Urban Walker (2001, 112–13). "Yet one of our recurrent ethical tasks is better suggested by the question 'What ought I to do *now*?' after the page is blotted or torn by our own or others' wrongdoing." It is this question that animates my project here. To that end, this book builds upon theories of reparative and restorative justice in political philosophy, feminist ethics, indigenous studies, and criminal justice, here extended into environmental contexts. In addition to relational accounts by Annette Baier, Robin Kimmerer, Janna Thompson, and Linda Radzik, my account of *reparative environmental justice* is most inspired by Walker's work on reparative justice and moral repair. Walker offers valuable theoretical tools for understanding and enacting what is needed to restore our moral relationships in the aftermath of injustice and wrongdoing, whether large or small. Where retribution metes out punishment, and restitution aims to make injured parties whole, "moral repair is the task of restoring or stabilizing—or in some cases creating—the basic elements that sustain human beings in a recognizably moral relationship" (Walker 2006a, 23). Reparation as Walker characterizes it is not really about monetary exchange or material restitution, popular connotations and associations notwithstanding, but the moral vulnerability of those who were wronged. Reparative justice is as much about victims' standing to demand accountability in their communities as it is about perpetrators' apologies or amends. For this reason, responsibilities of repair extend beyond specific parties to wrongdoing to include the moral communities of which they are members: "What demands repair is the state of the relationship between the victim and the wrongdoer, and among each and his or her community, that has been distorted, damaged, or destroyed" (Walker 2006b, 383).

Walker deliberately restricts her work on reparative justice to human conflict and wrongdoing narrowly framed. For reparative environmental justice to be applicable to ecological destruction and restoration, animal abuse and rehabilitation, or climate change requires non-anthropocentric and asynchronous revisions, which in turn raise significant theoretical and practical challenges. This book is therefore both an advertisement for the value of a relational-reparative approach to environmental justice and a critical assessment of the potential pitfalls attendant to it.

I am guided throughout by three interrelated questions. The first is primarily theoretical in its scope: for a variety of major debates in environmental ethics (the value of ecological restoration, intergenerational ethics, human-animal relationality, the amelioration of environmental injustice) how can reparative environmental justice provide a novel perspective? The second question is primarily practical: How might reparative environmental justice support some environmental practices and policies while enabling effective opposition to others? For example, what practical recommendations and critical appraisals can this approach underwrite for planned ecological restoration projects, sitings of locally undesirable land uses, climate agreements and adaptations, uses of traditional ecological knowledge, and animal sanctuaries or zoos? The third question is critically self-reflective: how might reparative environmental justice be ill-suited for extending relational repair into environmental ethics? The challenges to be articulated and assessed include issues of victim identification, interspecies trust, intergenerational forgiveness, asynchronous apology and amends, and wrongful repair. Many of these challenges are resolvable, or so I shall argue, but none can be easily dismissed.

These topics in their respective chapters are ordered such that each of them extends the model of reparative environmental justice: from environmental justice as reparative justice in chapter 2, relaxing the anthropocentric constraints with ecological restoration as moral repair in chapter 3, extending non-anthropocentric restorative justice further still with interspecies repair in chapter 4, grappling with asynchronous relational repair for climate change in chapter 5, and extending reparative environmental justice to epistemic injustices toward traditional ecological knowledge in chapter 6. In this way, each topic receives dedicated attention while the overarching project attempts to bring these topics together into a cohesive whole. Chapter 7 then applies the lessons articulated in these preceding chapters to the case of environmental restoration in the Chicago Wilderness, and chapter 8 concludes by reflecting on the strengths, weaknesses, and challenges for reparative environmental justice as a fruitful approach to environmental ethics.

This book was written in Chicago, on the traditional homelands of the Council of the Three Fires: the Ojibwe, Odawa, and Potawatomi Nations. Many other tribes such as the Ho-Chunk, Fox, Sac, Miami, and Menominee also call these lands home. Located at the confluences of the Chicago and Des Plaines rivers and along Lake Michigan, the region has long been a place for indigenous people to gather, trade, and maintain kinship ties. Chicago today is home to one of the largest urban American Indian communities in the United States, and the members of this community continue to contribute to the life of the city and celebrate their heritage, practice their traditions, and care for the land and waterways.[1]

I could not have written this book alone; I certainly did not write it alone. The arguments and ideas presented in these pages have benefited from questions, criticisms, and recommendations from many readers, conference participants, editors, and peer referees. Earlier versions of these chapters were given at the Conservation, Restoration, and Sustainability Conference at Brigham Young University in 2012, the Public Philosophy Conference at Emory University in 2013, the Central States Philosophical Association at Northwestern University in 2014, the International Society for Environmental Ethics Meeting in St. Louis in 2015, the Midwest Political Science Association in Chicago in 2016, the Feminist Epistemologies, Metaphysics, Methodologies & Science Studies Biennial Conference at University of Notre Dame in 2016, the North American Society for Social Philosophy at Loyola University–Chicago in 2017, the International Society for Environmental Ethics Summer Conference in central Oregon in 2019, the Philosophy of the City Conference at the University of Detroit-Mercy in 2019, and the First International Seminar on Restorative Justice and the Environment online in 2020. My thanks to the organizers, speakers, commentators, and other participants in these events.

The later stages of this project were written during my 2019–2020 Visiting Fellowship at the DePaul Humanities Center, where Peter Steeves and Anna Clissold made DePaul feel like home away from home. My thanks to Peter, Anna, the other DHC fellows, and others who participated in DHC workshops and events throughout my time at DePaul.

As a first-time author, I could not have asked for a better partner than Lexington Books. I am especially grateful to Jana Hodges-Kluck for her steadfast support for this project in its inception and to Holly Buchanan for her assistance in the final stages of its completion. My thanks also for a careful, thorough peer review process, which served to improve this book in many ways.

Chapter 3 is based in part on my 2017 article "Ecological Restorations as Practices of Moral Repair," published in *Ethics & the Environment*, and chapter 5 is based in part on my 2017 article "Climate Change and the Need for Intergenerational Reparative Justice," published in *Journal of Agricultural and Environmental Ethics*. I am grateful to anonymous reviewers for both journals, and to the journal editors for granting permission to reprint these works here.

I am also thankful for critical feedback on various parts of this project from many readers and interlocutors, including Monica Aufrecht, Sayoni Bose, Kimberly Dill, John Edward Davidson, Karen Emmerman, Juliana Oliveda Ferrari, Amy Flowerree, Rachel Fredericks, Sarah Fredericks, Karen Frost-Arnold, Cesar Cabezas Gamarra, Julia D. Gibson, David Golland, Ben Hale, Crystal Harris, Liam Heneghan, Marion Hourdequin, Eric Katz, Mike Konrath, Alexander Lee, Grazia Mannozzi, Samantha Noll, Kathryn Norlock, Novia Pagone, Chris Partridge, Jenny Partridge, Phil Partridge, Drew Peirce, Gaile Pohlhaus, Geoff Pynn, Jelena Radovic-Fanta, Baron Reed, Leonardo Rezende, Joao Salm, Carlos Santana, Jonathan Schwartz, Jarrod Shanahan, Ken Shockley, Bradley Smith, Allen Thompson, Joey Tuminello, Jessica Vogt, Ellie Walsh, Jaime Waters, Tama Weisman, Christopher White, John Yunger, and Jason Zingsheim.

Whatever else it accomplishes this book weaves together strands of a years-long conversation about human-ecological entanglements between two joyfully entangled people, a naturalist and a philosopher. Thank you for braiding this along with me, Negin joon, for holding the end as I pull and shift and weave, and letting me do the same for you.

NOTE

1. For more on the traditional homelands of the Council of the Three Fires and the American Indian community in Chicago today, see American Indian Center (2020).

Chapter 1

Justice after the Dam Breaks

INCIDENT AT CHURCH ROCK

Many Americans would be surprised to learn that the largest accidental release of radioactive material in US history was not the partial meltdown of Reactor 2 at Three Mile Island Nuclear Generating Station, nor any other nuclear plant, but rather a dam breach at the United Nuclear Corporation (UNC) uranium mill at Church Rock, bordering and encroaching disastrously on Navajo Nation tribal lands in northwestern New Mexico.[1] "About 5:30am on July 16, 1979, the Navajos of Church Rock chapter woke to the sound of running water, lots of it" (Pasternak 2010, 149). That night a 20-foot breach had opened in the earthen dam wall of a huge man-made waste pond and 93 million gallons of radioactive liquid—a slurry of nuclear tailings and water—poured into the adjacent arroyo and Rio Puerco riverbed. Livestock died, crops curdled along the riverbanks, and thousands of people were stranded without potable water. With the river contaminated, the Church Rock chapter of the Navajo Nation calculated that at least 30,000 gallons of water would be needed each day; United Nuclear distributed 600 gallon-bottles. When the Environmental Protection Agency's Superfund program was created in 1980, Pasternak (2010, 150) notes, "the off-reservation United Nuclear mill went on the list to be cleaned up, but the Rio Puerco did not."[2]

Despite its serious environmental impact, this case has seen little attention in environmental ethics. Perhaps this is because the ethical questions raised here are not much emphasized from an ideal-theoretic perspective. What UNC and the US government did at Church Rock was wrong, whether reckless, negligent, or otherwise understood. Ideally, this should have never happened. But it did happen, so what then?

All too often this is where environmental ethics drops away, either on the assumption that its work is done, or relatedly, that it has nothing further to usefully contribute. I want to resist both assumptions. Environmental ethics is not over after environmental wrongdoing. The ethical questions raised prior to acts of environmental damage and degradation are indeed important. Among the human and nonhuman entities affected, who or what deserves moral consideration, and why? Who or what has rights that have been respected, neglected, or violated? To the extent that something of value has been lost, what sort of value is it? How should we best characterize and conceptualize environmental harm, injustice, inequity, or wrongdoing otherwise construed? The tasks of anticipating and establishing the reality of environmental wrongdoing are difficult, yet we do ourselves a disservice if our ethical inquiry stops with them. Environmental ethics has a constructive role to play in guiding and making sense of life after environmental wrongdoing as well. (By *after* I mean that wrong has been done, but not necessarily that it has stopped.)

In this first chapter, I review several possible approaches to the moral aftermath of injustice and wrongs generally. I then consider how these various approaches apply to environmental contexts more specifically, on the way toward a preliminary sketch of the core concept of—and challenges for—reparative environmental justice.

AFTER INJUSTICE AND WRONGDOING

My goal in this book is not to offer a novel or definitive account of environmental wrongdoing. Many others have written extensively about the moral status of nonhuman animals, other living things and ecosystems, ideal distributions of resources and responsibilities, and human duties to one another and the rest of the biotic community. Indeed, much of this work will be relevant to the issues considered in this book and will be discussed along the way. But my main contention is that the aftermath of environmental injustice and other wrongdoings is itself a special context of ethical evaluation, one worth reckoning with on its own terms. For those who do not already regard animal suffering, global climate change, ecosystem destruction, cultural appropriation, or environmental injustice as contexts of wrongdoing, I doubt my assessments of these issues here will do much to change their minds.

Then again, sometimes an honest appraisal of the aftermath of wrongdoing can help us to acknowledge our role in the perpetration, allowance, and experience of such wrongdoing. Having theoretical and practical tools for responding constructively to environmental injustice and other wrongdoings can make the need for such a response less daunting to admit. Perhaps most importantly, people *already are* acting in the aftermath of environmental wrongdoing in more and less constructive ways. What I hope to accomplish is not only to show the importance of reparative justice in environmental contexts but also to shed some light on existing reparative practices, from ecological restoration and animal rehabilitation to intergenerational apology and reconciliation.

I think that it can be clarifying to begin with a particularly common response to wrongdoing, which is not to recognize it at all. This can be a lack of recognition by those who have committed the wrongdoing under consideration (perpetrators of wrongdoing), those against whom wrong has been committed (victims of wrongdoing[3]), or third-party community members. Something has gone wrong here, but what is it? Taking time to reflect on what is lacking when our response is no response at all can give us some direction, I think, for making sense of the substantive responses available to us. What's missing when we do not recognize wrongdoing? Asking this question invites the idea that, in addition to the initial wrongdoing itself, something further has gone wrong when we fail to respond properly to it.[4] Let's call this a second-order wrongdoing, not because it is less important than the initial wrongdoing but because of its contingency upon the initial wrongdoing. And if we can respond improperly in the aftermath of injustices and other wrongdoings, we can also respond properly. So understood, ameliorative justice is no less contingent than its negative counterpart. Both apply to nonideal circumstances.[5]

Perhaps what has gone wrong, beyond the initial wrong itself, is that a failure to recognize wrongs dooms us to repeat them. Consider the relatively happy timeline in which we actually do better in future instances than we have so far: our moral track record improves. Let's call this *amelioration as relative improvement*. Maybe we produce meaningfully fewer greenhouse gases than in years past; the populations of endangered species rebound; larger percentages of energy and other resources consumed come from renewable and sustainable sources; environmental policy decision-making is genuinely more democratic and inclusive than before. Few and far between are the line-items of our overall environmental track record that could not be improved upon. While a healthy dose of skepticism may be warranted toward governmental, corporate, and advocacy groups' claims of relative improvement on the environment, this kind of amelioration can and does happen. Indeed, the fact that it can be done is what grounds second-order criticisms of governments, corporations, and other institutions that fail to improve their prior environmental track records when and where improvement is possible.

Valuable as it can be, amelioration as relative improvement on its own would be inadequate as a response to wrongdoing. For one thing, relative improvement may not actually involve recognition of past practices *as* unjust, unfair, or otherwise wrong. We may make technical, social, or moral improvements *upon* the past without really learning *from* the past. So perhaps what is still missing even with relative improvements in our environmental track record is that crucial sense of moral understanding: we understand what was done wrong, how it was wrong, and our part in it, such that these things meaningfully inform our future analyses and behaviors. Let's call this *amelioration as iterative adjustment*. This second kind of amelioration can overlap and converge with amelioration as relative improvement, although neither is wholly reducible to the other. Sometimes we do better in the future *because* of lessons learned from the past, but not always so. Sometimes we get lucky, and sometimes moral understanding backfires at the level of overall ecological repair. But what is crucial here as a meaningful response to historical and persisting wrongdoings is that amelioration as iterative adjustment demands of us a recognition of wrongdoing that amelioration as relative improvement does not.

Yet amelioration as iterative adjustment also misses something important. Even as it requires our recognition of past practices as unjust, unfair, or otherwise wrongful, what does this sense of amelioration *do* about it? Iterative adjustment is backward-looking in terms of moral knowledge yet forward-looking in terms of moral application. I am tempted to say, as far as amelioration as iterative adjustment is concerned, that the moral lessons we learn can come from well-told fables instead of our actual morally fraught histories; we act differently because of what *we learn from* the past, not because of what *happened in* the past. This is not quite right, though. Sometimes the moral lessons toward iterative adjustment are timely warnings about our biases or prejudices, for example, which would not be as impactful without the recognition that these biases or prejudices are not only possible but real (and perhaps still persisting).

This example reminds us that something morally significant is missing if amelioration for past wrongdoing is cashed out entirely through the lessons we learn from and apply to future cases. We also need *amelioration as rectification*, something akin to Aristotle's corrective justice.[6] Contrary to distributive justice, which concerns how various goods ought to be distributed, corrective justice concerns that which is needed to restore conditions of justice damaged or destroyed by wrongdoing. Even if we learned from our past mistakes, have we done anything to correct them, or have we tried to put them behind us and pledged to do better going forward, to improve on distributive justice by learning from what our past and persisting moral failures can teach us? Sometimes this attitude makes sense: you cannot change the past, and there's no use crying over spilled milk. But to take this attitude not just toward mistakes or errors but toward injustices and other unethical practices would lose track of the specific normativity involved here. An injustice is not a bad piano recital, a subpar session at the gym, or a poorly cooked dinner. It is more like a nasty wound or a crack in the foundation that we ignore at our peril.

As before, amelioration as rectification can overlap and converge with iterative adjustment and relative improvement, but none of these are reducible to the others. It is possible to make progress on rectification without applying these lessons learned to moral improvements in future situations, and vice versa. What is distinctive about amelioration as rectification is its aim to set right what injustice has made wrong.[7]

RETRIBUTION, RESTITUTION, AND RELATIONAL REPAIR

Different approaches to rectification may be distinguished by their objects of emphasis. Consider responses to injustice and other wrongdoings focused primarily on the perpetrator(s) in question. *Retributive justice* (not to be confused with revenge[8]) is one such approach. Alex Walen (2014) identifies three principles at the heart of retributive justice: "(1) those who commit certain kinds of wrongful acts, paradigmatically serious crimes, morally deserve to suffer a proportionate punishment; (2) it is intrinsically morally good—good without reference to any other goods that might arise—if some legitimate punisher gives them the punishment they deserve; (3) it is morally impermissible intentionally to punish the innocent or to inflict disproportionately large punishments on wrong-doers."[9] To the extent that corrective justice happens after environmental harm, it tends to be retributive justice of some sort (jail times, fines, perhaps public shaming), and it does seem to be what many of us have in mind when we say we want perpetrators of serious environmental crimes to be held responsible for their actions.

It must be noted that not all perpetrator-centered approaches to corrective justice are focused on retribution; in addition to punishment, we might prioritize the goal of offenders' rehabilitation or redemption.[10] John Basl, for example, gives an account of the moral significance of ecological restoration that emphasizes its remediative potential. For ecological restoration as remediation, a person must identify the dispositions that led to their environmental wrongdoing, so that in doing the work of restoration, they may correctively cultivate more virtuous dispositions in themselves. As with restorationist work itself, cultivating virtuous dispositions is a process and not merely a desired outcome, requiring not only reflective contemplation but also sustained and deliberate activity. "Particular acts of restoration will be restitutive," Basl argues, "insofar as they are arranged in such a way that they are conducive to character remediation" (2010, 15).

I will not pretend to give a categorical refutation of retributive justice or perpetrator-centered responses to wrongdoing generally, but I want to draw attention to concerns especially relevant to the aftermath of environmental wrongdoing. For one, it gives little direction for cases in which identifiable perpetrators cannot be found, yet the need for corrective justice remains. What is missing when perpetrators die or become otherwise unavailable for punishment? We could say the remnant is regret that an opportunity for retribution or rehabilitation has been lost; but for many of us, this is at best an incomplete explanation, one that marginalizes victims and damages caused by the injustice at hand, damages beyond perpetrators' own moral characters. Another shortcoming, then, is that retribution and even rehabilitation can be divorced from the needs and priorities of those who were injured. Those who are punished may or may not acknowledge their wrongdoing, and those who were injured may or may not have a voice in the retributive process, which may or may not align with what injured parties actually need.[11]

Perpetrators of injustice are surely relevant to its correction, but perhaps centering on them is misplaced; perhaps then, our efforts to set things right again should focus instead on the material conditions in place prior to injustice displacing them. *Restitutive justice*[12] is less concerned with punishment than with remedy, returning things how they were prior to injustice. As Desmond Tutu argued, "If you take my pen and say you are sorry but don't give me the pen back, nothing has happened."[13] Restitution seems an especially natural response to theft and fraud, though even here things can get complicated. What does someone like Charles Ponzi or Bernie Madoff *owe* their defrauded investors: restitution of their original investments, the figures promised as returns on that investment, or perhaps something somewhere in between?[14] Putting such complications aside, there is something incomplete about the aim of restoring material conditions as a response to wrongdoing. If one person accidentally takes my pen and another steals it, they should both return the pen; but doesn't the latter owe me something more than the former, something that differentiates honest error from theft?

The material conditions disrupted by injustice are certainly relevant to amelioration, but especially for serious historical injustices, is putting things back as they once were an adequate response? When the wrongdoing in question cannot be reduced to money or commodities like a television or a box of cereal, literal restitution might be impossible. Maybe you can return my pen, but you can't unscramble an egg, put the toothpaste back in the tube, or remove radiation from sheep and people exposed to negligently irradiated water. In such cases, restitution is more a matter of compensating for harm done rather than undoing the harm in question.[15]

Among prominent environmental ethicists, Paul Taylor is notable for directly addressing the issue of restitution given human exploitation of the natural world for our ends. Taylor describes this restitution in terms of compensation, such that restitutive justice means "bringing about an amount of good that is comparable (as far as can be reasonably estimated) to the amount of evil to be compensated for" (1986, 306). In her discussion of animal sanctuaries as sites of remedy and repair, Karen Emmerman (2014a) argues that this analysis of restitution as compensation has real problems. One is that treating evil as something that can be compensated invites a dangerous sort of satisfaction or complacency: that when all is accounted for, in the final moral reckoning, perhaps there was no wrongdoing after all. One might respond that comparable compensation is an ideal never actually to be reached, a benchmark by which to judge partial compensations as more or less satisfactory, and recognizing this limitation we will not be tempted to see injustice and other wrongdoings as ever completely expunged. Even still, those with Emmerman might argue that compensation frames our response to environmental wrongdoing in the wrong sort of way and so gives the wrong set of benchmarks by which to evaluate our attempted responses.[16]

If retribution focuses on perpetrators' offenses and punishments they deserve, and restitution on victims' losses and making them whole again, *restorative justice* and *reparative justice* focus primarily on the relationships between victims, offenders, and their communities, relationships damaged or destroyed by wrongdoing.[17] "Moral repair," as Margaret Walker (2006a, 23) puts it, "is the task of restoring or stabilizing—or in some cases creating—the basic elements that sustain human beings in a recognizably moral relationship." The focus is on restoring the conditions for relationships, which means "*restoring or creating trust and hope in a shared sense of value and responsibility*" (Walker 2006a, 28). This puts considerable importance on our acknowledgments of wrongdoing, apologies, amends, and eventually forgiveness. Responsibilities of repair apply to the individuals directly involved but also the communities of which they are part. In redressing wrongs, moral repair values relational restoration over retribution or compensation: "restoration or construction of confidence, trust, and hope in the reality of shared moral standards and of our reliability in meeting and enforcing them" (Walker 2001, 120).

Despite popular assumptions, reparative justice is not really about monetary payments. "The fundamental issue in reparations," Walker (2010, 15) says, "is the moral vulnerability of victims of serious wrongs."[18] If you want to do right by us afterward, it matters how you hurt us, whether you stole a car, destroyed a garden, or irradiated a river. But for proponents and practitioners of reparative or restorative justice, as Rupert Ross says, "the harm done is only peripherally about 'stuff.' Instead, the harm is understood in the *relational* realm" (2006, xvii). Reparative justice and material compensation are crucially different. "When reparations are paid, they are often token,"

Neumann and Thompson (2015, 9) explain, "more a symbol than full compensation for harms done." Walker (2015a, 133) agrees that reparations are indeed symbolic, yet she warns us not to read this as rendering memorials, apologies, and acts of amends as somehow less real for their symbolic value. When we seek to make amends as part of a reparative process, we do not offer proportionate compensation for wrongdoing or even an apologetically partial compensation for wrongdoing, because the goal is not repayment but rather relational repair, and what victims need to see happen in order to rebuild trust and, eventually, extend forgiveness.[19] "Reparations are a medium for the contentious yet hopeful negotiation in the present of proper recognition of the past and proper terms of relation in the future" (Walker 2015b, 217).

Understood in this way, reparative justice does not ask us to look backward but to do what we can to repair moral relationships and communities that have been hurt by historical and ongoing injustices. Admitting and apologizing for our part in wrongdoing and making amends matter because they do work of repair. The key to reparative justice is communication between those who commit injustice and those injured by it. Amends are not meant as charity or partial compensation for what was lost. They work because of the "expressive burden" (Walker 2013) they carry: their ability to convey regret, acknowledgments of wrongdoing, and most of all, recognition of those we have hurt as members of our shared moral community deserving of equal respect and consideration. "Part of what is involved in rectifying an injustice is an acknowledgement on the part of the transgressor that what he is doing is required of him because of his prior error," Bernard Boxill (1972, 118) argues. "Without that acknowledgement of error, the injurer implies that the injured has been treated in a manner than befits him; he cannot feel that the injured party is his equal."[20] What is missing or at least incomplete about a punitive response to injustices and other wrongdoings concerns its ability to carry that expressive burden. Whether carceral or financial, punishments may be warranted, and even hold individuals, groups, and institutions accountable, and yet fail to make any real progress toward restoring or stabilizing healthy trust relationships in the aftermath of injustice.

A note about terminology: rectification, restitution, restoration, reparation, and reconciliation are all closely related, but as we consider the variety of normative responses to the aftermath of injustice and other wrongdoings, a little precision might be useful. To *rectify* is to put right what injustice has put wrong. As I have discussed in this section, this general corrective aim admits of more specific versions. *Restitution* is one way to attempt to put things right, whether by returning things to their rightful owners or instead retroactively compensating victims for their losses. (One could be compensated in advance for anticipated harms or losses, of course, but this isn't usually what we mean by restitution.) I follow Walker in understanding *reparation* in terms of relational repair, such that what is being set right after wrongdoing are the conditions of trust, hope, and confidence in mutually recognized and defensible shared standards[21] between the relevant parties. *Restorative justice* and reparative justice are closely related, perhaps more of a disciplinary distinction (the former comes out of criminal-justice scholarship and practices, the latter is more common for international law and historical injustices) than a substantive one. Along the way I will borrow liberally from both literatures, but as a matter of emphasis I want to maintain focus on repairing relationships. Finally, while *reconciliation* is certainly close to relational repair, one need not entail the other. Not all reconciliations are morally healthy, and working toward repair of the conditions for morally healthy relationships does not always require or recommend reconciling with those who have wronged us. What matters most is not that we must come back together, but that we rebuild the conditions capable of underwriting the mutual trust, hope, and accountability that make morally healthy reconciliation possible.[22]

REPARATIVE ENVIRONMENTAL JUSTICE AND ITS CHALLENGES

Environmental philosophies concerning our moral obligations to each other and the natural world too rarely address the aftermath of environmental destruction and degradation. Ideally, we would never do ourselves, one another, other animals, or biotic systems wrong—but given that we do, we need theoretical and practical resources relevant and applicable to environmental contexts. I propose reparative environmental justice as one such nonideal approach to environmental ethics, building upon and extending the accounts of reparative and restorative justice introduced in the previous section.

Let's begin with the idea that environmental injustices and other environmental wrongdoings do moral damage to our relationships, which thus calls for a process of moral repair if and when these relationships are to become morally healthier than they have been rendered in the aftermath of wrongdoing. Wherever else it might lead, taking this as our starting point draws our attention to the relational dimensions of environmental ethics. I see this project as situated alongside and joyfully indebted to relational approaches to environmentalism taken by such scholars as Robin Kimmerer and Marion Hourdequin (to whom we will return in chapter 3 on restoration ecology) and Lori Gruen and Karen Emmerman (more on them in chapter 4 on interspecies repair).[23] Reparative environmental justice requires us to take seriously the importance of relationality in environmental ethics. That said, I think there is something relevant here for many environmental philosophies. I do not claim that all environmental wrongdoing is best understood relationally or that there are no wrongful acts without victims.[24] My priority here is not to establish reparative environmental justice as a comprehensive environmental ethic, but to show how it might serve as a constructive guide for making sense of the aftermath of wrongdoing in various environmental contexts.

As Linda Radzik explains, wrongdoing does moral damage to our relationships in many ways, including victims' and perpetrators' relationships with each other, with third-party community members, and even themselves. When we are victimized, we may wonder whether we ourselves are somehow responsible, doubt the accuracy of our own perceptions of injustice and judgments of others' moral character, or internalize a sense of ourselves as morally inferior. When we do wrong, we may feel disgusted with or alienated from ourselves, or perhaps even worse, develop a distorted sense of our morally superiority (Radzik 2009; 78; see also Tan 2007, 286). We might add also that wrongdoing damages our relationships by undermining our ability to successfully collaborate with others: perhaps because we can no longer trust each other, because we actively distrust each other, or because our continued trust rests on a morally rotten foundation (Baier 1986). Unrepaired wrongdoing can also damage relationships between victims and third-party community members when the former are unable to move the latter to recognize their unresolved grievances. Unrepaired wrongdoing can damage relationships between perpetrators and third parties when the latter are unable to hold the latter accountable for their wrongdoing, which over time has a corrosive moral influence on these relationships and the community as a whole.

It would strain credibility to claim that every environmental wrongdoing does moral damage in all of these ways. Let us instead take a lesson from the feminist tradition of care ethics, which urges us to appreciate the moral significance of our relationships but does not presume to treat all such relationships the same. As we consider environmental relationality more broadly, this basic lesson becomes even more important: that we stand in many different kinds of relationship with the many varied members of what Aldo Leopold (1966) calls our *biotic community*. There is no one blueprint for understanding relational damage or for enacting relational repair that applies uniformly across all of our environmentally significant relationships.[25]

Walker (1998, 5–10) and Radzik (2009, 200) both deliberately restrict their accounts of moral repair to human conflicts and wrongdoing narrowly construed, and in doing so they are not alone among those who write on the aftermath of harm.[26] For reparative environmental justice to be applicable to ecological destruction and restoration, animal exploitation and rehabilitation, and global climate change will require non-anthropocentric and asynchronous revision. This in turn raises theoretical and practical challenges including issues of victim identification, interspecies and intergenerational trust and forgiveness, unrepentant perpetrators, preplanned amends, and the possibility of wrongful repair. Each of these challenges receives more detailed critical attention in the coming chapters; here I simply aim to introduce and discuss them briefly.

Challenges of Victim Identification

"In restorative justice, what demands repair is a state of the relationship between the victim and the wrongdoer, and among each and her community, that has been distorted, damaged, or destroyed" (Walker 2006b, 383). Building on Walker's insight here means attending to *victims* and *relationships* as they pertain to our histories of environmental damage and domination. Yet some may wonder about the difficulties in identifying victims of environmental wrongdoing and whether the goal of restoring trust, hope, and accountability rightly extends into environmental contexts.

One challenge concerns the identification of injured parties *as* victims. Erinn Gilson (2011) and Haalboom and Natcher (2012) warn that for so-called vulnerable populations, "vulnerable" and "victim" can be perilous labels for those to whom they are attached. Such appellations can obscure victims' particular social contexts, their distinctive forms of knowledge, and their own complicity in environmental degradation, and more generally, serve to disempower victims by putting the onus for action in the moral aftermath of environmental wrongdoing on perpetrators. Whether the language of victimhood serves to disempower is a serious challenge not easily dismissed. I think extending Walker's prioritization of victims' subjectivity can serve as a much needed safeguard against their marginalization in the planning, implementation, and continuation of environmental reparations. Keeping the focus on victims of environmental wrongdoing as the proper beneficiaries of moral repair provides clarification and direction. Rather than expediency or generic public utility, practices of reparative environmental justice so conceived are organized around the needs, priorities, and experiences of those wronged in environmental contexts.

This prioritization only works if environmental damage and degradation have actual victims, which is Markku Oksanen's core criticism of Bill Jordan on ecology restoration as reparation. "If we consider nature as a victim of our wrongful actions, we lack the victim," Oksanen (2008, 105) writes. "Nature lacks victimhood, that is, the essential psychological and complex qualities that characterize the life and the experiences of human victims and the nature of human communality." Though their positive accounts differ considerably, John Basl (2010) and Lee et al. (2014) both put aside thorny questions of victim identification in their respective discussions of the value and importance of ecological restoration. While I do agree that victim identification presents nontrivial theoretical and practical challenges, perhaps a healthy dose of environmental-ethical pluralism might help here. Different environmental philosophies assign different moral standing to people, animals, and ecosystems, yet each still assess environmental destruction as harmful, whether to ecosystems directly, individual creatures negatively impacted, or human beings either directly or indirectly. The core idea of reparative environmental justice thus holds, explicated differently for different ethics in terms of the victims and relevant moral relationships identifies. The pluralist suggestion is that those motivated by different underlying environmental philosophies might still work capably in coalition on projects of environmental moral repair. As with any coalition, overlapping consensus may well be fragile, and our different versions of reparative environmental justice will not always give the same ecological or ethical recommendations. Still a contingent, coalitional, open-ended approach seems a promising place from which to start.

Finally, the asynchronistic nature of climate justice presents challenges for the prioritization of victim preferences and subjectivities. To the extent that amends are made by the members of earlier generations and victims are the members of later generations, the former may be unable to access the latter's subjectivities to direct their reparative work. Or we might substitute our own subjectivities through understandable if misplaced empathy, giving the impression that contrite wrongdoers have deferred to victims' preferences and perspectives while the victims' actual subjectivity remain inaccessible. I suggest that the inaccessibility of future victims' subjectivity be incorporated in reparative climate justice as a guiding methodological insight and caution for epistemic humility. This means pursuing open-ended reparative practices that allow for and even invite future persons' revisions as they see fit.

Challenges of Environmental Trust and Forgiveness

Loosening both the anthropocentric and synchronous constraints typical of restorative and reparative justice presents further challenges for the roles of trust and forgiveness in reparative environmental justice. Of particular concern are environmental ethics that assign significant moral standing to individual nonhuman animals, species, and ecosystems but stop well short of ascribing capacities for trust and forgiveness to them. To riff off a query from Jeremy Bentham often raised by proponents of animal rights and liberation, the question is not "Can they trust?" nor "Can they forgive?" but "Can they suffer?" Sentience as a moral criterion overlaps with, but is not equivalent to capacities for forgiveness and trust. How can animal-rights ethics make sense of reparative environmental justice for nonhuman victims of human abuse and exploitation? Similarly, how are we to understand the role of trust and forgiveness on holistic environmental ethics that identify entire ecosystems or biotic communities as the central objects of moral consideration?

Even restricting our attention to human relationships, the intergenerational nature of global climate change challenges some basic assumptions about what victim-offender forgiveness and reconciliation look like. For one thing, if past generations are recognized as victims of climate injustice, the fact of causal asymmetry would seem to preclude past persons' forgiveness of present or future persons as part of the repair process. For an earlier generation to forgive a later generation, it would seem that their forgiveness must be wholly articulated in advance, without the chance to respond to later generations' amends or their demonstrated trustworthiness. A second part of this challenge concerns the forgiveness (or lack thereof) of future people. We could jettison the requirements of forgiveness and renewed trust in asynchronous cases, but as later chapters explore in more detail, perhaps the better option is to keep victims' forgiveness at the heart of intergenerational reparative environmental justice while giving up the expectation that perpetrators ought to be able to know whether they have been forgiven and that renewed trustworthiness has been established to victims' actual satisfaction.

Challenges of Prospective Amends

If reparative environmental justice is meant to offer a constructive response to environmental wrongdoing, is there a role for *prospective* or *preplanned* apologies or amends as part of morally legitimate relational repair? At the very least there is certainly something peculiar about making amends *before* wrongdoing. We might suspect that preplanned reparations are made in bad faith, or further still, make for a performative contradiction: can I sincerely apologize for something that I nonetheless proceed to do? If we think of environmental amelioration first and foremost in terms of compensation, then the challenge here is not so great: after all, we routinely compensate each other for anticipated losses. Indeed, the widespread practice of environmental restoration as preplanned mitigation for future ecosystem destruction or degradation (revisited in more detail in chapter 3) seems to follow this logic, trading acre for acre to justify future development projects. But part of the power of reparative environmental justice is its ability to underwrite principled opposition to preplanned mitigation projects (or so I shall argue), so the challenge of prospective or preplanned amends cannot be too quickly dismissed.

Another version of this challenge arises for intergenerational environmental wrongdoing such as climate change (more on this in chapter 5), given the considerable lag between anthropogenic causes and their felt climatological effects. Rather than victims experiencing the harmful effects of perpetrators' climate injustices and *then* experiencing some restitution or reparation, here the restitution or reparation is made by perpetrators in the morally unusual space after said injustices are committed but before they are experienced. Does the asynchronicity of reparative climate justice heighten the contradiction of prospective amends or rather resolve it? Here as elsewhere, the devil is in the details.

Perpetrators and Practitioners of Repair

If there is a real challenge of *when* the work of environmental moral repair is done, there is a similar challenge of *who* exactly does this work. If reparative environmental justice is meant to be a victim-centered process and not merely a victim-centered outcome, then genuinely contrite wrongdoers seeking forgiveness and renewed trust and accountability must *do the work*, literally. "Restorative justice seeks to build and strengthen individuals' and communities' capacities to do justice actively," Walker (2006b, 383) reminds us, "and not to surrender the role of doing justice to experts, professionals, or 'the state,' which should play facilitating roles." Now sometimes the reparative work is done by the perpetrators of previous harm or destruction, such as farmers who care for the very animals they have harmed or those who remove drainage tiles and do prescribed burns and reseeding to restore their properties to ecologically appropriate prairies and wetlands. For ecological restoration, animal sanctuaries, and climate adaptation organized and conducted today, however, professionalization is the norm. Restoration technicians and expert caretakers do the actual day-to-day work, while those directly responsible for the harm in question might be footing the bill but do very little reparative work. Can such work really be the work of amends if not done by the main perpetrators of environmental wrongdoing, and where does this leave those professionals and volunteers who do the weeding, seeding, feeding, and other caretaking?

The Very Goal of Relational Repair

Is reparative environmental justice always an appropriate goal, given historical and persisting environmental injustices and other serious environmental wrongdoings? Here relational repair might be criticized either on the grounds that it might conflict with other vital measures toward environmental justice or because some relationships just should not be repaired. The relational approach to feminist ethics in which Walker's work and my own project are situated has long been attentive to the moral significance of our particular care relationships in ways that other moral theories have not. One such insight is that many of our most important relationships and attendant responsibilities are not ones we deliberately, freely choose in the way that, say, two independent equal parties might freely agree to a contract (Baier 1986; Gruen 2015, 64). But attending to our chosen and unchosen relationships does not thereby obligate us to stay in these relationships, forever and always. It does not contradict reparative justice to recognize that the moral conditions for our relationships sometimes cannot or should not be repaired.

Yet abolitionism is more viable for some relationships than others, particularly when grave ecological conditions are at stake. As Annette Baier (1981, 178) warns, "This power to end the human community's existence could justifiably be exercised only in conditions so extreme that one could sincerely believe that past generations would concur in the judgment that it all should end." This sets a high bar, one befitting the relational-reparative approach to environmental ethics explored throughout this book.

NOTES

1. As Eric Jantz, a lawyer from the New Mexico Environmental Law Center, told *Vice* in August 2019, "The Church Rock spill symbolizes the governmental and societal indifference to the impacts of uranium development on indigenous lands. The Church Rock spill is the third largest nuclear accident after Fukushima and Chernobyl, and the largest in the US in terms of radiation released, but nobody knows about it" (Gilbert 2019).

2. See also Wasserman et al. (1982), Hungate (2005), Brugge et al. (2007), and Gilbert (2019).

3. I use the language of victims and victimhood advisedly. As will be discussed in later chapters, concerns have been raised about how victim appellations can be disempowering or can obscure the ways in which so-called victims have played active roles in the very injustices at hand. For my purposes, "victim" is meant simply to refer to those who have been wronged. Like Margaret Walker, it is important to me to be able to prioritize and center victim subjectivities in making sense of the aftermath of injustice. This is compatible with the possibility that multiple parties might be wronged at once, even as their subjectivities converge or diverge in various ways. It allows for the possibility that victims of injustice are also perpetrators of injustice, including the initial or second-order injustices. But we should make note of the fact that a victim-centered approach can tend to ignore or marginalize cases of wrongdoing that have no victims, cases in which no moral patient has been done wrong.

4. Smith (2008, 65) reminds us that "failing to apologize for injuring someone can actually be more harmful than the injury itself." On the ethics of remembrance for historical injustices, Blustein (2015, 84) argues that "the presumption should be that forgetting victims and their suffering is an additional wrong on top of the initial one." Or as Biggar (2001, 10) puts it, "To suffer an injury and have it ignored is to be told, effectively, 'What happens to you doesn't matter, because you don't matter.'"

5. On the relationship between ideal theory and injustice, see also Mills (2005) and Jaggar (2019).

6. Aristotle (1990, Book V); on distributive justice and corrective justice, see Waldron (1992), Thompson (2015), and Butt (2015). Walker (2015b, 217) seems to treat corrective justice and compensatory or restitutive justice as synonymous, rather than the latter as a type of the former. To avoid confusion, I will refer to amelioration as rectification as the more general category, in which reparative justice and restitutive justice are more specific alternatives.

7. As Radzik (2004, 145) puts it, "Wrongs are 'righted' in the sense that ships are 'righted.' To right a ship is to restore its balance. To right a wrong is to restore the social balance, to bring the relationship or the community back into harmony."

8. Nozick (1981, 367); see also Jacoby (1983), Moore (1987), Minow (1998), and Klimchuk (2001).

9. For more on principles of retributive justice, see also Cottingham (1979) and Walker (1999).

10. This is not to assume punishment and rehabilitation or redemption must be mutually exclusive: see Hampton (1984) on rehabilitation and moral education and Radzik (2009) for a perpetrator-centered theory of reconciliation.

11. "Victims find that they are mere footnotes in the process we call justice," writes Howard Zehr (2013, 23). "If they are involved in their case at all, it will likely be as witnesses; if the state does not need them as witnesses, they will not be part of their own case. The offender has taken power from them and now, instead of returning power to them, the criminal law system also denies them power." For a similar critique of retributive justice, see Radzik (2009, 45).

12. On restitutive theories of punishment, see Holmgren (1983), Hershenov (1999), and Barnett (2013).

13. As quoted in Berlinger (2005, 61).

14. On Ponzi schemes, see Zuckoff (2005), Markopolos (2010), Arvedlund (2010), Henriques (2011), and Almassi (2018).

15. See Holmgren (1983, 37) and Butt (2015, 178).

16. For further criticism of restitution, see Radzik (2009, 47–48).

17. On reparations and reparative justice, see Boxill (1972), Wheeler (1997), Thompson (2002), D'Costa (2005), de Grieff (2006), Renar (2006), Sepinwall (2006), Walker (2006b, 2010, 2015b), and Thompson (2015).

18. Torpey (2015, 65) writes, "The terms 'reparation' and 'reparations' have a variety of meanings in international law and human rights parlance, only a relatively small proportion of which have to do with monetary compensation. This fact is consistent with the claim that is frequently made by those seeking reparations that 'it is not about the money.' For many such claimants, the noneconomic forms of reparation would suffice."

19. As Walker explains, this is where Bass's (2012) characterization of reparations as a "noble lie" misses the mark. "No one is more acutely aware than victims of grave wrongs that there is no undoing many of them; what they seek as reparations can hardly be that" (Walker 2015b, 212).

20. See also Murphy and Hampton (1988) and Wilk (2017).

21. Walker (2015a, 134). Janna Thompson seems to refer to reparative justice and rectification interchangeably, such that "restitution for historical wrongs belongs to the general subject of reparative justice" (2015, 46). I diverge from Thompson and align with Walker on terminology, not because one usage is clearly more correct than the other, but because I want to distinguish between approaches to rectification that emphasize compensation and those that emphasize relational repair.

22. As Walker puts it, "Moral repair requires more than reviving capacities for trust and hope in wronged and seared souls; it requires good reasons to think that a society is once again worthy of trust and hope" (2001, 122). That said, given the range of accounts for each, the distinctions among restorative justice, reparative justice, and reconciliation are fuzzy. On reconciliation, see Tutu (1999), Biggar (2001), Schaap (2005), Radzik (2009), Verdeja (2009), Murphy (2010), Murdock (2016), Emerick (2017), McGregor (2018), Murdock (2018b), and Whyte (2018).

23. On relational accounts of environmentalism, see Hourdequin and Wong (2005), Gruen (2009), Kimmerer (2011), Hourdequin (2012), Kimmerer (2013), Emmerman (2014a, 2014b), Gruen (2015), and Emmerman (2019).

24. Here I would echo Radzik (2009, 78–79): "Not everything that is wrong with wrongdoing can be reduced to the category of relational harm. The advantage of thinking of wrongdoing as primarily damaging relationships is that it encourages us to attend to all of the parties who are negatively affected by wrongdoing and to identify the various kinds of harms that wrongs might cause."

25. "Although an appreciation of the complex relations we are in with the more-than-sensate world has important implications, not all relations are ethically equivalent" (Gruen 2015, 69)

26. Murdock (2016, 1) argues that "the presentation of ecological relations and political relations as separate and oppositional to each other leads to similarly bifurcated methods of harm analysis and proposed remedies."

Chapter 2

Environmental Injustice and Its Amelioration

Environmental ethicists and political theorists have written on environmental-justice movements and restorative-justice practices, respectively, but intersections and areas of convergence between them have received fairly little philosophical attention. This chapter considers what a relational-reparative approach has to offer environmental-justice scholarship and activism, and at the same time, how environmental justice can temper tendencies toward anthropocentrism and romanticism in restorative justice. Reparative and environmental justice are compatible, and complementary, without being redundant; environmental justice can benefit significantly from reparative justice, and vice versa.

TOXIC WASTES IN WARREN COUNTY AND BEYOND

The former Ward Transformer Company facility in Raleigh NC is now a Superfund site, added to the EPA's National Priorities List in 2003 for its high levels of soil, sediment, and fish-tissue contamination (US EPA 2019). But in 1978, the facility was in full operation, and President Buck Ward had a disposal problem: thousands of gallons of oil contaminated with polychlorinated biphenyls (PCBs) used to keep transformers from overheating. In years past Ward had disposed of PCB-contaminated waste oil fairly easily, whether at a landfill or a sewage treatment plant, but with passage of the 1976 Toxic Substances Control Act, PCB manufacture, use, and disposal had become more tightly regulated and expensive (McGurty 2007, 24). Legal disposal was now permitted at just two licensed incineration sites in Missouri and New Jersey. As Bruce Siceloff (2014) put it in the *Raleigh News & Observer*, "Ward Transformer found a cheaper alternative."

Ward hired Robert Burns's waste-hauling company, and for nearly two weeks under the cover of night Burns tanker trucks illegally dumped roughly 30,000 gallons of PCB-contaminated oil along 200 miles of rural roads across a dozen counties in North Carolina. Burns would spend five years in prison, and Ward served nine months and was held liable for the cost of the cleanup.[1] This story does not end with legal actions against Ward and Burns, however, because important though they might otherwise be, incarceration and monetary penalties cannot turn back the clock. Miles of highway were still contaminated; tons of toxic soil would have to be removed and a containment site found. From an ideal-theoretic perspective, such egregious environmental degradation should not have happened. There is no question that what Ward and Burns did was wrong. But in the aftermath of their wrongful actions, who then must bear the environmental burdens so produced?

One potential course of action was to move the contaminated soil out of state to an existing toxic-waste facility in Alabama. A local option was to excavate and contain soil at smaller sites near the affected areas. North Carolina opted instead to create a wholly new landfill to which the contaminated soil from across the state would be sent and contained on a former farm in Afton, Warren County (McGurty 2007, 50–60). The opposition from Warren County residents was swift and tenacious, but after three years of legal battles, still unsuccessful. Despite concerns about the negative health and ecological impacts of PCB containment on the Afton site,[2] the NIMBY label on Warren County activism seemed fitting and hard to shake. Eileen McGurty explains:

> The NIMBY label imposed by the authorities implied that the landfill opponents' parochialism was harmful because the landfill was a societal necessity. In the early stages of the opposition, the locals had no way of countering this claim. . . . The parochial view of the problem ultimately undermined the objective of stopping the landfill but was able to sustain collective action for four years and prepare the county for a new and more expansive collective action frame of environmental racism. (McGurty 2007, 53)

A landfill dedicated to toxic wastes like PCB-contaminated soil is undeniably an environmental burden. The counterargument against landfill opponents is not to deny this, but to insist that the burden is unavoidable. What an environmental-justice analysis brings that NIMBYism cannot is, among other things, an emphasis on the demographic distribution of environmental burdens and benefits. Who enjoys more or less of the benefits of resource extraction and consumption, and who must shoulder more or less of the attendant risks and other related burdens?[3]

The residents of Warren County did not produce the PCBs in question, nor did they especially benefit from their usage, nor were they responsible for their illegal dumping. And while Burns tanker trucks released PCB-contaminated oil along many miles of Warren County roadsides, the total environmental degradation was hardly limited to that one county, so even putting aside the question of responsibility for environmental damage, the cleanup was not simply a local problem for which the Afton landfill would be a local solution. In this case many communities across the state had the same problem of illegally dumped toxic waste, yet only one was being made to take on the environmental and health impacts of its solution—and it was no coincidence, the activists argued, that Warren County was one of the poorest counties with one of the highest percentages of African American residents in the state (Geider and Waneck 1983, 50).

In the United States, the environmental-justice movement in which the Warren County protests figured prominently was explicitly positioned as an alternative to mainstream environmentalism, rather than an outgrowth from it (Taylor 2000; Sze and London 2008). Activists in Warren County and across the nation in the 1980s were more likely to see their struggles as part of the civil-rights movement than existing environmental movements, more aligned with the Southern Christian Leadership Conference and United Church of Christ than Greenpeace or the Sierra Club.[4] This is not to say mainstream environmental groups at the time were unconcerned about toxic waste or pollution—far from it. But this concern was rarely if ever conveyed or pursued in terms of distributive justice. Consider *An Environmental Agenda for the Future*, a 1985 declaration of shared environmentalist priorities cosigned by the "Group of Ten" mainstream environmental organizations,[5] addressing issues including population growth, nuclear energy, water, pollution, and urban issues. "The *Agenda* approached each of the issues with two assumptions," McGurty points out (2007, 126): that "environmental problems impacted everyone equally and solutions advocated in the *Agenda* would protect everyone to the same degree." At best we could say that these organizations failed to recognize the inequitable environmental burdens shouldered by poor people and people of color; at worst their campaigns actively produced or exacerbated inequities. As Richard Moore and other leaders of the Southwest Organizing Project (SWOP) put it in their open letter to the Group of Ten, "Your organizations continue to support and promote policies that emphasize the cleanup and preservation of the environment on the backs of working people in general and people of color in particular."[6]

The Sierra Club's North Carolina chapter and the Conservation Council of North Carolina had been involved in early legal battles over the Warren County landfill, concerned as they were about potential water contamination and other environmental damages from the PCB-laced soil to be brought and contained there (McGurty 2007, 105). But neither of these mainstream groups participated in the direct-action protests that began in September 1982, after the legal battles had concluded and trucks began hauling tons of contaminated soil to the landfill. In keeping with the civil-rights alignment of the environmental-justice movement, hundreds of people marched from Coley Springs Baptist Church to meet the trucks and block their entrance. In the months to come, approximately 7,000 truckloads brought 40,000 tons of contaminated soil to the Afton landfill and 523 predominantly African American protestors were arrested there. Included among them was Benjamin Chavis, executive director of the United Church of Christ Commission for Racial Justice and future NAACP executive director and organizer of the Million Man March. Chavis also coined the phrase *environmental racism*, meaning "racial discrimination in the deliberated targeting of ethnic and minority communities for exposure to toxic and hazardous waste sites and facilities, coupled with the systematic exclusion of minorities in environmental policy making, enforcement, and remediation" (Lazarus 2000, 259).

"The PCB protest failed to prevent the landfill from being completed," Geider and Waneck (1983, 52) wrote, "but it succeeded in a number of ways." Locally, activists secured agreements that no more landfills would be built in Warren County and water levels in and around the Afton landfill would be closely monitored; when groundwater contamination was indeed discovered in the early 1990s, these same activists were able to demand a seat at the table in the detoxification process (McGurty 2007, 148). Nationally, Warren County became a rallying cry for the growing movement. The protests outside the Afton landfill in September 1982 spurred lawmakers in the Congressional Black Caucus to request a General Accounting Office investigation of disparities in toxic-waste locations in the South.[7] The Government Accountability Office (GAO) findings confirmed that Warren County was not unique: poor African American communities across the region were disproportionately impacted by hazardous waste sites. But since the GAO had deliberately limited its study to Southern states, nationwide research was still needed.

"Toxic Wastes and Race in the United States" was published by the United Church of Christ to provide just this sort of credible evidence, which campaigns against toxic waste in majority-minority and predominately poor communities across the country could cite in making the case that some Americans bear inequitable burdens compared to others. Authored by Chavis and Charles Lee, the study was published in 1987 to widespread recognition. In years thereafter, other studies would come to similar findings on race, class, and environmental inequity, but the UCC study is still often cited in popular and scholarly accounts of environmental justice.[8] Esme Murdock explains it thusly:

The significance of this study should not be understated, especially considering the societal dismissals that often discount, silence, and erase the knowledge(s), ways of knowing, and testimony of marginalized and vulnerable populations. This report made legible the realities, which are largely rendered invisible and isolated within these communities, precisely because the experiences of these communities generated the investigation.[9]

The study's primary finding was the identification of race—even more than wealth or income—as the strongest indicator of toxic-waste facility proximity in the United States. This result was shown by comparing demographic statistics of those zip codes where one or more hazardous waste sites are located with zip codes without any such sites. Racial minority groups were overrepresented in areas containing hazardous waste facilities, nationwide as well as in major cities. For example, "Blacks were overrepresented in the populations of metropolitan areas with the largest number of uncontrolled toxic waste sites" (like Memphis and Chicago) and "half of Asian/Pacific Islanders and Native Americans live in communities with uncontrolled toxic waste sites" (Chavis and Lee 1987). In 2007 UCC published a follow-up study, "Toxic Waste and Race at Twenty," which improved on the prior study's methodology, looking at 2-kilometer circles around toxic-waste facilities rather than zip codes, which had invited criticism that facilities located at the corners and borders of designated areas may produce misleading data on geographic proximity. With methodological improvements and more recent data to work with, the new study concluded that "people of color are found to be more concentrated around hazardous waste facilities than previously shown" (Bullard et al. 2007).[10]

Momentum built with the UCC study, the SWOP letter, Robert Bullard's *Dumping in Dixie*, and the First National People of Color Environmental Leadership Summit where hundreds of activists gathered to draft and adopt seventeen Principles of Environmental Justice, which have served as a cornerstone for environmental-justice work since.[11] By the early 1990s, the discourse and values of environmental justice began to see uptake from mainstream institutions. The Sierra Club, Natural Resources Defense Council, and Greenpeace added environmental justice to their organizational platforms, the Environmental Protection Agency (EPA) opened its Office of Environmental Equity (later renamed the Office of Environmental Justice) and US president Bill Clinton issued Executive Order 12898, directing federal agencies to take environmental justice into account in their work.[12] Such wider uptake certainly seems like progress, and yet institutional appropriations of environmental justice is not without some cause for concern, especially for grassroots activists wary of watered-down versions of environmental justice and fair-weather commitments that last only until the next administration or other changes in organizational leadership.[13] "It's very important that the environmental agenda we develop is one developed *with* communities of color," urged Vivien Li, chair of the Sierra Club Ethnic Diversity Task Force, in a 1993 roundtable discussion on environmental justice. "Not imposed on them, but rather forged together in the spirit of mutual respect and trust" ("A Place at the Table" 1993, 90).[14]

AFTER ENVIRONMENTAL INJUSTICE

Environmental justice continues to be a powerful frame for both activists and scholars in the twenty-first century reckoning with social inequities, environmental policies, and related phenomena. The scope, applications, and methodologies of environmental justice have broadened to include food justice, indigenous justice, climate change, and migration, among other global issues, as well as quantitative, qualitative, and interdisciplinary approaches (Mohai et al. 2009; Schlosberg 2013; Agyeman et al. 2016). One thing that has been fairly consistent about environmental justice over the years, as Mohai et al. (2009) note, is that most attention in activism and scholarship goes to establishing the fact of environmental injustice, racism, or inequity, whether in specific cases or as a general phenomenon. I would not say this emphasis is misplaced: for activists, establishing a community's environmental-justice status is significant under EPA rules and regulation, and for scholars, identifying and interrogating the methods by which claims of environmental injustice, racism, and inequity can be established is likewise important. This is true whether the injustice at hand concerns inequitable toxic exposure, lack of recognition, exclusion from participation, or a combination of these.

But what comes after? One reason why environmental-justice work can be so controversial is "that it is not immediately obvious what should be done after an injustice has been documented," Mohai et al. (2009, 407) write. "It seems that societies that only reluctantly admit inequality and racial injustice have been hesitant to develop plans to solve the problem."[15] What is amelioration for environmental injustice, once the facts of injustice have been established? Recall these three senses of amelioration:

- *amelioration as relative improvement*—doing better compared to how we have done so far;
- *amelioration as iterative adjustment*—revising future activities in light of prior failures;
- *amelioration as rectification*—restoring conditions of justice damaged by prior failures.

We need not necessarily choose among these senses of amelioration, but neither do they always converge. Relentlessly forward-looking improvement can be achieved without learning lessons from past failures, and even a reflective approach to iterative adjustment may not serve to rectify previous wrongs. We might also differentiate among restitutive, retributive, and reparative forms of corrective justice or rectification, the first focusing on victims' losses and making them whole again, the second on perpetrators' offenses and the punishments they deserve, and the third on repairing moral communities and relationships damaged by injustice or other wrongdoing.

When the aftermath of environmental injustice is addressed, it tends to be in terms of iterative adjustment, improvement, compensation, or retribution: recall the prison sentences for Burns and Ward, and the latter's financial liability for the PCB cleanup. Financial and carceral punishment for environmental damage have their place but cannot be our complete account of environmental justice, as Chavis argued in the aforementioned roundtable discussion for *Sierra Club Magazine*. "When the movement first got going, I think some whites actually became afraid, because they thought it was a movement of retribution," he said. "It is not a movement of retribution—it is a movement for justice" ("A Place at the Table" 1993, 52). What sort of justice? Chavis explains that environmental-justice activists are not trying to move toxic-waste facilities out of black communities and into white ones— "we're saying they should not be in anyone's community" (ibid.).

This is an inclusive and unifying aspirational ideal. But as amelioration for already existing environmental injustices it doesn't offer much direction for corrective action. Consider also Peter Wenz's (2001) LULU points proposal, meant to counterbalance inequitable siting of toxic wastes and other locally undesirable land uses (LULUs) in some communities more than others. Unlike Chavis's aspirational ideal, Wenz allows that LULUs must be located in someone's community, while still aspiring for roughly equal distribution across communities regardless of race, class, or other demographic differences. Whatever its strengths, the LULU points proposal is amelioration in the sense of improvement (as future points-based siting decisions begin to even out LULU geographies) or perhaps iterative adjustment (given that the proposal itself is built on the lesson of past and present environmental inequities) but it is not corrective justice. At worst the LULU points schema frames past and present inequities as only conditionally unjust, no longer wrong once sitings have been evened out; at best past and present inequities continue to be recognized as environmental injustices, but Wenz's proposal provides no conceptual or practical resources toward their rectification.

Stella Capek argues that while compensation has a place in environmental justice, for many communities, "getting satisfactory compensation is the most difficult struggle of all" (1993, 19). Capek carefully traces a case of environmental injustice in the Carver Terrace neighborhood of Texarkana TX, where "residents' ability to mobilize effectively for social change was intimately linked to their gradual adoption of an 'environmental justice' frame" (1993, 6). Carver Terrace was declared a Superfund site in 1984. For many years before its residential redevelopment, the site was used for commercial production and disposal of wood-treatment chemicals. In the early 1960s production ceased, buildings were removed, and the site was rezoned for residential use, the first lots sold in 1967. Capek notes that, at a time when most Texarkana neighborhoods still had racially restrictive covenants, home ownership in Carver Terrace was an incredibly desirable opportunity for middle-class African Americans, and the revelation of Carver Terrace's toxicity was an unwelcome shock. When the EPA recommended a "soil-washing" technique to clean the neighborhood without relocating residents, wary residents mobilized around relocation instead.

The Carver Terrace environmental-justice campaign built momentum and national recognition throughout the 1980s, and in November 1990 President George Bush signed an appropriation bill allocating five million dollars for Carver Terrace buyouts and relocations. Yet even for a relative success story like this, compensation as a corrective response to injustice still presents problems. Residents were to be compensated for the value of their homes, which often was not enough to purchase another home elsewhere. The buyout and relocation process was slow and complicated, adding years of further exposure for residents well beyond the 1990 appropriations bill, on top of their exposures after the 1984 Superfund declaration and indeterminate levels of exposure since housing was built in 1967. And when buyouts and relocations were finally complete, a close-knit neighborhood was now scattered across the state. "In the end," Capek says, "there is no adequate compensation for the loss of a functioning community" (1993, 20).[16]

Collin and Collin champion environmental reparations as a bridge to sustainability and equity, "both spiritual and environmental medicine for healing and reconciliation" (2005, 221). They are explicit that reparations in the aftermath of environmental racism and injustice cannot simply be monetary awards nor achieved by creating more parks. They propose creation of *environmental protection districts*, modeled on the existing legal designation of historic preservation districts. Their argument continues as follows:

> The urgency of the need to repair the most impacted places on earth is based not simply on claims of justice, but on recognition of the common dependence of all living things on heavily affected living systems. . . . Whether the urgently needed environmental reparations in urban communities of color should be accompanied by an apology or by acknowledgement of harm done, is an issue that demands attention, but the urgency itself is not debatable. (Collin and Collin 2005, 217)

I propose that the case for environmental reparations made by Collin and Collin might be further strengthened by theories of restorative and reparative justice. In the next section, we'll consider the theoretical and practical advantages of putting reparative environmental justice into practice in the aftermath of environmental injustice.

EJ + RJ: CONVERGENCE AND COMPLEMENTARITY

How can reparative justice provide guidance for environmental justice; how can environmental justice bolster reparative and restorative accounts of justice? Let us begin by considering some key points of convergence between them.

Both attend to the nonideal. Reparative justice is based in nonideal circumstances. Ideally, we'd never do each other wrong, but given that we have, we need theoretical and practical tools to respond to injustice and wrongdoing generally. Though there are exceptions (see Wenz 1988), environmental-justice advocates also tend to focus on nonideal cases, concerned with actually existing inequities more than exemplars of justice. For both reparative and environmental justice, then, the default object of analysis is not success; instead we begin with failure and work from there.[17] *Given* injustice, how can perpetrators, victims, and their communities go forward? How can we repair relationships in the aftermath of injustice, not just ignore it and try to do better next time?

Both have grassroots origins followed by mainstream and institutional recognition. As noted earlier, the rise of environmental justice was more an alternative to mainstream environmentalism than an extension of it. Eventually the values and discourse of environmental justice began to see uptake from mainstream groups and institutions, but the balance is still precarious. Restorative justice likewise has had an uneasy relationship with institutional recognition. Restorative values and practices have long been recognized as foundational for some communities' justice systems. "From an indigenous perspective, justice must transcend the distributive, capitalist model," Dina Gilio-Whitaker (2019, 26) argues. "Indigenous modes of justice typically reflect a restorative orientation."[18] For other communities, restorative justice offers a local alternative to the police, courts, and prisons—and yet the success of restorative justice has also seen its incorporation into existing, punitive systems of criminal justice and law enforcement, where we find police officers assisting in restorative justice and judges using it in sentencing.[19] On the one hand, institutional appreciation for restorative justice seems like progress; on the other, grassroots practitioners and scholars of restorative justice worry about the limitations of restorative justice within mainstream criminal-justice systems and institutions.[20]

Both prioritize victims' subjectivities and participation. One key feature of restorative justice identified by proponents and practitioners is its "openly and actively participatory" (Woolford 2009, 14) nature.[21] Repair is a deliberate and engaged process and not merely a desired outcome, which requires wrongdoers who seek renewed trust and forgiveness to do the work of making amends. Restorative justice builds our capacities to *do* justice, not just delegate the work to be done by experts, professionals, or the state (Walker 2006b, 383).[22] Understanding restorative justice this way means taking seriously the need for the parties to precipitating injustices to actively participate in processes of amelioration (Woolford 2009, 90; Johnstone 2002, 64).

Here then is a significant point of complementarity: *restorative and reparative approaches help reaffirm the importance of environmental justice beyond distributive justice*. Particularly when taken up as an internal critique of mainstream environmentalism, environmental justice is often described mainly if not entirely in terms of distributive justice, that we should care not just about environment destruction, degradation, and harms in the aggregate but more specifically about who tends to shoulder more environmental burdens and who tends to enjoy the benefits made possible. Distributive questions are certainly important, but the danger is taking issues of distributive justice to capture the full extent of environmental justice, such that success is entirely a matter of achieving equitable distributions of hazardous waste and other undesirable land uses. "The earliest academic reflections on environmental justice originally focused on the existence of inequity in the distribution of environmental bads," writes David Schlosberg (2013, 38). "Yet for all the focus on the reality of these inequities, environmental injustice was never *only* about such maldistributions." Many activists and scholars have a pluralist conception of environmental injustice, where it also is about recognition and participation, including recognition of the values that marginalized groups are allowed or denied to bring into their participation in environmental policy-making.[23] Thus environmental justice is about people being able to "speak for ourselves" (Agyeman et al. 2009) as much as it is about distributions of environmental burdens and benefits. "The community is transformed by the grassroots environmental justice groups established in the midst of environmental struggles," write Cole and Foster. "These groups help transform marginal communities from passive victims to significant actors in environmental decision-making processes" (2011, 14).

Collin and Collin also argue that environmental justice must have a participatory dimension, that environmental reparations require not only dedicated land but also "additional community capacity building," such that those wronged by environmental inequity have not only the right but the ability to participate in environmental decision-making (2005, 220). I suggest that their proposal can be bolstered by reparative justice's explanation of the need for acknowledgment and apology. *Restorative and reparative approaches reaffirm the importance of environmental justice beyond compensation.* "Part of what is involved in rectifying an injustice," Boxill (1972, 118) reminds us, "is an acknowledgement on the part of the transgressor that what he is doing is required of him because of his prior error." For perpetrators, accepting responsibility for our actions and recognizing them as wrongful may not be sufficient, but these minimal tasks are surely necessary to even start to set things right with victims of injustice. For our actions to constitute amends we must acknowledge our wrongdoing and offer our actions as redress for it. "Without that acknowledgement, reparative actions are charitable, compassionate, or generous, even dutifully so," Walker (2006a, 191) says, "but they do not 'make amends.'" Taken together, apologies and amends show that the perpetrator and larger community recognize victims' moral standing to call for accountability in the face of injustice.

Just what sort of amends are called for will be determined by what victims need to repair their strained relationships and communities, toward building trust and eventually forgiveness. Practically speaking, there may or may not be much overlap in what would be required as compensation for environmental injustice and what is required for relational repair. But this, of course, will be an open question, depending on what particular victims need to repair particular relationships in light of the particular environmental injustices under consideration. Recall again Carver Terrace, where residents and their allies secured relocation and buyouts for homeowners, despite the EPA's initial recommendation of soil-washing as an adequate cleanup response. That residents' perspectives on amelioration were (eventually) prioritized over expert advice is encouraging from a reparative environmental justice approach. Because individual homeowners were left to secure buyouts and relocations as individuals rather than as members of a damaged community, any sense of community repair was inevitably marginalized if not entirely lost.

Last but not least, *environmental justice can temper tendencies of romanticism in restorative justice.* Skepticism on restorative justice takes many forms. Proponents of punitive and carceral forms of criminal justice, for example, may dismiss restorative justice and reparations as well-meaning but naïve. A criticism coming from a different direction is that restorative justice is predicated on an overly sanguine view of the past, of a morally and socially just prior state of affairs to be restored. Proponents of transformative justice who break from restorative justice, for example, do so because they see it as too backward-looking and not sufficiently alive to the need for systemic transformation. Rather than trying to restore something that (perhaps never really) was, they would have us build something *new*, something *better* in the aftermath of injustice, and what must be rebuilt are not only human relationships but also unjust social-economic systems.[24] An environmental-justice framework bolsters the need for systemic critique, since establishing and addressing distributive and social inequities necessarily calls for large-scale comparative assessments. As with other civil-rights campaigns, many environmental-justice activists are more interested in where justice might take us from here, rather than getting back to a more innocent past.

RECOGNITION AND (NON-)ANTHROPOCENTRISM

Environmental justice challenges mainstream environmental priorities: in the primacy given to wilderness preservation, for example, and debates over the intrinsic value of nature (Schlosberg 2013, 39). On an environmental-justice perspective, the environment isn't something untouched by human beings; rather, "The environment is where we live, where we work, where we play, and where we learn" (Lee 1996, 6–7; Cole and Foster 2001, 16). Given such contrasts, it can be easy to read environmental justice as anthropocentric; George Sessions's (1995) highly critical reading of environmental justice as social justice posing as environmentalism is predicated on just such an interpretation. Certainly much attention has been given to environmental inequities between and among different groups of people, extending back to the GAO and UCC studies. Meanwhile the Principles of Environmental Justice adopted at the First National People of Color Environmental Leadership Summit begin by affirming "the sacredness of Mother Earth, ecological unity and interdependence of all species, and the right to be free from ecological destruction."[25]

I think it's fair to say that some approaches to environmental justice are more human-centered than others, whether articulated in explicitly anthropocentric terms or implicitly attending more to human inequities while neglecting nonhuman interests. This is one reason for ambivalence on the institutionalization of environmental justice by the EPA and other such institutions, which inevitably operationalize environmental-justice values in anthropocentric terms. For those who take seriously the non-anthropocentrism of the Principles of Environmental Justice, adoption of restorative-justice principles and practices may raise similar worries, since restorative justice taken up by Western judicial systems and scholars has often been explicitly or presumptively operationalized as human-centered as well.[26]

The question of anthropocentrism seems to see similarly ambivalent answers in restorative justice and environmental justice: explicitly non-anthropocentric versions have been articulated and enacted, yet mainstream, institutionalized versions are often anthropocentric. If conceptual and normative priority is given to victims' subjectivities, and human subjectivities are easier to reliably access than nonhuman ones, does this mean reparative environmental justice inevitably reinforces a presumptive anthropocentrism? If participation and procedural justice are important for reparative justice and environmental justice, is this limited to human participation?

Despite these concerns, there is reason to expect that taking the participatory requirements of reparative environmental justice seriously for historically marginalized and environmentally burdened human groups will undercut a presumption of anthropocentrism rather than reinforce it. Not all humans are human-centered in their approaches to environmental decision-making.[27] As Robert Figueroa, Kyle Whyte, Deborah McGregor, and others have argued,[28] it is not enough that marginalized and excluded groups be allowed in the door: *recognition justice* concerns not just our participation, but also the values we are allowed or prohibited to bring into our participation. "Recognition justice does not privilege non-tribal discourses of environmental justice and seek to find ways to fit tribes in," Whyte (2011, 203) explains. "Rather, recognition justice gives priority to the host of radically different ideas of justice that arise from different cultures and memories." Taking relational repair seriously means taking recognition justice seriously, since mainstream delegitimization and misrepresentation of tribal and other environmental philosophies is among the many historical and persisting injustices in need of amelioration.

RESISTANCE, RECONCILIATION, AND REPAIR

The reparative response to environmental justice as I have outlined it here is a process aimed at relational and community repair, toward rebuilding the moral conditions necessary for healthy relationships and moral accountability. But what if relational repair between perpetrators and their victims is a misplaced priority? Some genuinely regretful perpetrators of injustice and wrongdoing might wish to repair their relationships with those they have wronged, but other perpetrators would seem not to care much about it. And if we do prioritize victim subjectivity, whether relational repair is the right goal should not be determined by what (some) perpetrators want, but what victims by their own lights need—in which case, what communities repeatedly and systematically subjected to environmental wrongdoing need might not be repair but rather resistance.

Resistance may seem ill-fitting in other contexts in which restorative justice is applied, where individual perpetrators have harmed individual victims in discrete acts of wrongdoing. It may well be an open question whether victims are ready and willing to sit down with their attackers, let alone forgive them; indeed, one concern about institutionalized forms of restorative justice is that reparation becomes preplanned, routine, removed from victims' control, and systematized in a judicial bureaucracy. Nevertheless wariness to forgive or enter into a reparative process is not quite the same thing as resistance, which is collective political action taken by the comparatively powerless against the comparatively powerful. Those victims and perpetrators who participate in restorative-justice processes might not stand in that sort of relationship; victims and perpetrators might both be fairly powerless. Meanwhile power dynamics are central to environmental justice, whether they concern government-government, government-community, corporate-community, or a combination of these relationships.[29] Mobilization of marginalized groups in solidarity and resistance against powerful corporate and governmental actors may seem to be more reasonable than relational repair.

I can offer two partial answers to this question of resistance and repair in the aftermath of environmental injustice, the first of which comes from an environmental-justice framework which recognizes and reaffirms ecological interconnectedness and relationality. This is one place where grounding of environmental justice in US civil-rights movements converges at least in part with mainstream environmentalism, where Martin Luther King Jr.'s "inescapable network of mutuality, tied in a single garment of destiny" (1964) aligns with Aldo Leopold's "thinking like a mountain" (1966). As Collin and Collin put the point, "There is no 'separate but equal' in nature, no 'separate but equal' way to solve the issue of sustainability. There are no allowable sacrifice zones, human or otherwise, in our ecological inter-connectedness, and there is no exit" (2005, 212). We continue to be ecologically interrelated, even when we have morally dysfunctional relationships. Relational repair cannot be entirely dismissed if only because separatism is not possible.

But this doesn't tell us anything about the *timeline* of repair, how soon reparative processes should begin or end, which as with other aspects of reparative justice should be guided by the priorities and perspectives of those who have been wronged rather than perpetrators' desires for closure and forgiveness. Kyle Whyte explains how the call for political reconciliation between indigenous people and settler nations directed by the latter against the former—and presuming the latter's perspective on injustice and its appropriate amelioration—can be worse than nothing. Not merely ineffective and oblivious, settler attempts at reconciliation can actually recapitulate settler colonialism itself, another way of constructing an illusory moral high ground from which to explain away and justify historical and continuing injustice. In this way, Whyte (2018, 287) argues, "The illusion engenders a system that is resilient *against* reconciliation."[30]

The second response, then, is to recognize resistance as a precondition for reparative justice: resistance now may be necessary for repair later. If reparative justice is about community repair as well as specific victim-perpetrator relationships, then when perpetrators are unrepentant, or victims are unwilling or unable to rebuild trust with perpetrators, resistance may be a crucial part of reasserting and reaffirming victims' standing to demand accountability after wrongdoing. As Winona LaDuke argues, resistance can be needed in order to reassert and reaffirm the nature and requirements of specific damaged relationships. "We are not part of and do not wish to be part of the mainstream of America. We are different," she says. "America has to come to terms with our difference" ("A Place at the Table" 1993, 55). This is not to deny that indigenous Americans and mainstream Americans are interconnected, ecologically or otherwise; but whatever else happens, relational repair is impossible until the latter recognize that the nature of this relationship is not simply whatever we take it to be.

ENVIRONMENTALIST, HEAL THYSELF

I would like to conclude by highlighting a specific context for reparative environmental justice, namely, in the relationship between mainstream environmental organizations and environmental-justice activists. It is surely a good thing when mainstream environmentalism comes to recognize and even adopt environmental-justice values and priorities. But if and when this happens, at least two reparative concerns remain.

The first is a tendency by mainstream environmental groups to embrace environmental justice in a relentlessly forward-looking way: amelioration as relative improvement, and in some cases amelioration as iterative adjustment, but comparatively little amelioration as repair. Mainstream groups recognizing environmental justice in the 1990s were more likely to affirm its value than to acknowledge their culpability for not having done so before. To the extent acknowledgments are made, they tend to be broad admissions of fallibility (we could or should have done better) than taking responsibility for harms to those excluded or marginalized from their organizational priorities. When mainstream environmentalists do recognize their own neglect of environmental justice as itself an environmental injustice, the emphasis is on lessons learned and changes going forward. But relational repair demands more than this; it demands at the very least some level of accountability and amends toward reconciliation.

Dina Gilio-Whitaker's book on indigenous environmental justice takes its title *As Long as Grass Grows* from US president James Monroe's 1817 speech to the Cherokee. "You are now in a country where you can be happy; no white man shall ever again disturb you," Monroe asserted. "As long as water flows, or grass grows upon the earth, or the sun rises to show your pathway, or you kindle your camp fires, so long shall you be protected from your present habitations" (Gilio-Whitaker 2019, 15). The second concern about mainstream environmentalists' appropriations of environmental-justice values and priorities, then, is the trustworthiness and sustainability of such self-avowed commitments. This is a familiar problem for any alliance, but particularly with allies who must prove themselves against a long track record of silence, complicity, and commission of injustice. Can you trust your self-avowed allies to stick with it?[31]

In "Whither the Eco-Warrior?," Florence Williams (1997) discusses criticisms by Greenpeace International that Greenpeace USA "hasn't been doing enough direct action, that is has become distracted by its recent focus on small local issues and environmental-justice campaigns." 1997 saw a sea change of leadership: Earth First! cofounder Mike Roselle joined the Greenpeace USA board of directors in January as Winona LaDuke and Ron Daniels—the only members of color and both strong proponents of environmental-justice work—left in September. "Greenpeace is not about community organizing or going door-to-door. It's about raising hell and getting things done. I was trying to get this organization back on track, and I think we've done it," said Roselle. "'They forced me out,' says LaDuke, an Anishinabekwe Indian who bitterly regrets the gutting of campaigns against pesticides, dioxins, and PCBs, many of which pollute poor communities." The shift in organizational priorities was apparent. Longtime Greenpeace USA campaigner Jack Weinberg offered this assessment:

> There were a handful of people of color who were at the forefront of our environmental-justice work whose jobs have been eliminated, and that is a matter of grave urgency to many on the staff. . . . I don't think it was intentional that these people were targeted, but it could cause irreparable harm to our work in some communities. (Williams 1997)

Reparative justice requires more than just perpetrator acknowledgment and apology for past wrongdoings, important though they are. Victim-centered amends as reparative practices serve not only to communicate perpetrators' regrets and apologies but also to repair accountability and trustful collaboration.

The good news for mainstream environmentalists who genuinely wish to pursue relational repair is that we do not have to guess at how to make amends. At the general level, reparative environmental justice directs participants in relational repair to prioritize the perspectives and priorities of those who have been wronged; at the particular level, environmental-justice activist groups have repeatedly given mainstream environmental groups not only their criticism but also accompanying counsel on ameliorative actions to take toward reconciliation.[32] The example of Greenpeace's fair-weather commitment to environmental justice is, I think, usefully contrasted with the perspective offered by Leslie Fields, longtime director of environmental justice for the Sierra Club. "Our prime directive is to work at the community's request," Fields says of Sierra Club environmental-justice projects. "And these initiatives are created by the communities and the local Sierra Club chapters" (Durlin 2010).[33] While there is no quick fix for relational repair, this community-first approach points us in the right sort of direction.

NOTES

1. *United States v. Ward* (1985), McGurty (2007, 167), Siceloff (2014), and Taylor (2014, 18–19).
2. See *Warren County v. North Carolina* (1981), Geider and Waneck (1983, 51), and Taylor (2014, 15–17).
3. See Capek (1993), Sandweiss (1998), and Taylor (2000).
4. See Bullard (1994), Taylor (2000), Cole and Foster (2001), and Mohai et al. (2009). "The Sierra Club became the archetypal environmental organization challenged by environmental justice advocates, who thought the club focused on nature, not people" (McGurty 2007, 126.)
5. These were the Sierra Club, Environmental Defense Fund, Nature Conservancy, Greenpeace, National Audubon Society, Natural Resources Defense Council, National Wildlife Federation, World Wildlife Fund, Defenders of Wildlife, and the Wilderness Society (see Cahn 1985).
6. Quoted in "A Place at the Table" (1993, 54); see also Durlin (2010).
7. McGurty (2007, 114); US General Accounting Office (1983).
8. See Bullard (1994), Cole and Foster (2001), Merchant (2003), and Cutter (2006). "This study was a landmark; all subsequent research about equity in environmental risk either refined, substantiated, or rejected the UCC's methodology or conclusion" (McGurty 2007, 116).
9. Murdock (2019, 306).
10. See also Been (1992), Mohai (1994), Goldman and Fitton (1994), Been (1995), Crawford (1996), Bowen (2002), Steel and Whyte (2012), and Almassi (2016).
11. "Some say that from October 21-22, 1991, the environmental movement in the United States changed forever" (Cole and Foster 2001, 32). See also Bullard (1993, 1994), Durlin (2010), and First National People of Color Environmental Leadership Summit (1991).
12. Durlin (2010, 19–21) and Mohai et al. (2009, 409–10).
13. Williams (1997) and Mohai et al. (2009, 421). "Without a doubt, it is at the level of local community struggles that the Environmental Justice Movement has had its clearest victories" (Brulle and Pellow 2006, 113). Meanwhile, as Bullard puts it, "The mission of the federal EPA was never designed to address environmental policies and practices that result in unfair, unjust, and inequitable outcomes. The EPA and other governmental officials are not likely to ask the questions that go to the heart of environmental justice" (2001, 155).
14. Machlis (1990, 278) warns, "The central irony of conservation in the democratic regime may be that sometimes conservation groups rob power instead of give power, and thus resemble the architects of dominion and environmental disregard."

15. Fredericks (2011, 63) notes that policies for environmental justice are too rarely accompanied by mechanisms for monitoring progress toward environmental justice; to that end, she offers "a framework to aid in the development of indices that link multiple environmental justice issues within and across spatial and temporal scales."

16. See also Figueroa and Mills (2001, 435–36) for further critical discussion of compensation as a response to environmental injustice.

17. In this way, both are more reminiscent of Judith Shklar's (1990) failure-first political theory than, say, John Rawls's (1971) ideal theory of justice. On the other hand, it should be noted that when Bullard (2001, 154) describes environmental justice as taking "a public health model of prevention (elimination of the threat before harm occurs) as the preferred strategy," he is closer to Rawls than Shklar in their respective approaches.

18. See also Consedine (1995) and Ross (2006).

19. See Zehr (1990), Sullivan and Tifft (2006, 5), and Woolford (2009, 12).

20. On the relationship between restorative justice and criminal justice, see Pavlich (2005) and Daly (2003, 2006).

21. See also Young (1983), Figueroa and Mills (2001), and Whyte (2011).

22. "*Restorative justice is a process*: it does not follow a pre-set linear course; instead, it is a process through which the parties are offered an opportunity to create new meanings out of a situation that they, in most cases, experienced as negative. Given this processual character, it is impossible to impose specific desired outcomes on restorative justice" (Woolford 2009, 16).

23. See Warren (1999), Holifield (2001), Schlosberg (2002), Schlosberg (2007), Sze and London (2008), McGregor (2009), and Whyte (2011). The EPA (1992) defines environmental justice as requiring the meaningful involvement of all people in the development, implementation, and enforcement of environmental rules and regulations.

24. Morris (2000); see also Braithwaite (1999), Lofton (2004), Nocella (2011), Zehr (2011), Acorn (2012), and Robinson (2013).

25. First National People of Color Leadership Summit (1991). Other principles also affirm the moral standing of nonhuman life. As Schlosberg (2013, 44) observes, "Most of the discussion is about environmental bads and injustices to human beings, but the origins of environmental injustices are as much in the treatment of the nonhuman realm as in relations among human beings."

26. See Zehr (1990), Walker (1998), Johnstone (2002), and Woolford (2009).

27. See for example LaDuke (1999), McGregor (2009), and Roy (2017).

28. See Figueroa (2006), Figueroa and Waitt (2008), McGregor (2009), Figueroa and Waitt (2010), and Whyte (2011).

29. See Capek (1993), McGregor (2009), and Whyte (2011).

30. See Cook (2016), Jung (2018), McGregor (2018), and Murdock (2018a, 2018b).

31. McKenzie (2014) and Greenberg (2014).

32. See Moore's remarks on mainstream environmental groups ("A Place at the Table" 1993, 53).

33. See also Figueroa (2001, 176).

Chapter 3

A Relational Revaluation of Ecological Restoration

"Today, ecological restoration offers the potential to recover lands and waters from environmental damages that societies inflict on Earth's natural resources by misuse or mismanagement," observe Margaret Palmer and coauthors in their preface to *Foundations of Restoration Ecology* (2016, xiii). Restoration ecology differs from both conservation biology and land development (which of course differ from each other) in that restoration finds direction and meaning in the aftermath of environmental degradation and destruction—what Hettinger (2012, 39) calls the "essentially regrettable character" of ecological restoration. And yet part of the appeal of restoration is that it offers us a constructive response to environmental harm with tangible, measurable results.

The value of ecological restoration has seen extensive criticism and defense in environmental ethics over the past thirty years. Proponents like William Jordan (2003) emphasize the human and ecological benefits of restoration,[1] while critics like Eric Katz (1996) and Freya Matthews (1999) have worried that restoration can be deceptive, delusional, and domineering.[2] As ethical debates on ecological restoration developed, pragmatists like Andrew Light (2006) contributed a welcome perspective, urging philosophers who study restoration not to overlook its public-facing dimensions.[3] Most recently, John Basl (2010), Rohwer and Marris (2016), and others have further reoriented the debate by making sense of ecological restoration in terms of moral restitution and responsibility, such that the ends of restoration are not only ecosystem configurations but moral matters themselves.[4]

Here I aim to complement such efforts to reorient the debate by drawing our attention to the theoretical and practical merits of approaching and implementing restoration projects as practices of reparative environmental justice. Recognizing restoration projects as amends for ecological destruction and degradation can further strengthen their normative basis.

Ecological restoration projects at their best are warranted on ecological, instrumental, and aesthetic grounds; publicly organized and implemented restorations can have meaningful social and cultural benefits too.[5] Alongside appeals to the total utility of restoring an ecosystem, reparative environmental justice reframes restoration work, including why, how, and by whom it is done, as practices of apology, acknowledgment, and amends for historical and ongoing environmental harm. In what follows, then, I consider how a relational-reparative account of ecological restoration can usefully disrupt ethical debates on restoration while also underwriting criticism of restoration projects undertaken today as standardized, professionalized, and preplanned forms of mitigation.

RESTORATION IN RELATION

For four years my academic home in northern Illinois was a suburban college abutting a locally important ecological site: a small farm for most of the last century, over the past thirty years it has been restored (or perhaps transformed) into a flourishing savanna with prairie grasses, oaks, and wetlands. Years of deliberate destruction and construction have created the Rollins Savanna, "including disabling over thirteen miles of drainage tiles, installing over 200,000 native wetland plants, 2,000 native trees and shrubs, and several hundred acres of prairie restoration" (Cannon 2007, 1) through the steady labor of professional restoration technicians and public volunteers, including students from my environmental ethics courses and me. Weeding, cutting, stacking, and seeding, we came to appreciate restoration ecology as an active, engaged, and yet far from deterministic practice.

The Rollins Savanna is ecologically and socially significant, but it is not unique, not even in northern Illinois. William Stevens's *Miracle under the Oaks* (1996) tells the story of the Chicago Wilderness, a governmental-grassroots environmental alliance pursuing conservation, education, and restoration initiatives that still continue today. Ecological degradation and destruction in the Chicago Wilderness area stretching across Illinois, Indiana, Michigan, and Wisconsin has taken many forms including habitat degradation, invasive-species introduction, water and soil pollution, and the destruction of prairies and savannas for agricultural, industrial, and residential purposes. Stephanie Mills describes the challenge of ecological restoration in the Chicago Wilderness this way:

> The midwestern tallgrass prairie all but vanished, overturned by the steel plow in favor of the relentless monotony of agriculture, which amounts to planting lots of the same damn thing—corn in Illinois, usually—and then attempting

to protect it from all the plagues that even-aged monocultures are prey to. Sodbusters and years later the advance of suburbia made wild fires, and hence the prairie and oak openings than depended on them, even more a thing of the past. Benign neglect will not suffice to keep these previous few remnant ecosystems alive. Fires must be lit and skillfully managed, and many other tasks performed as well. (Mills 1995, 131)

Ecological restoration is not limited to seeding, prescribed burns, invasive-species removal, and native-species repopulation, but also extends to restoration of lakes, creeks, wetlands, rivers, and other displaced and degraded waterways (Shore 2011). Application of restoration-ecology theory to ecological-restoration practice (Palmer et al. 2016, 7) takes many forms for different projects across different settings. "There is a tremendous range of restoration activity—many hundreds of projects across the United States and abroad in a variety of ecosystems, reefs, salt marshes, arid farmlands, prairie pot-holes, Alpine meadows, mangrove swamps," Mills (1995, 3) reminds us; "The list is about as long as the list of the kinds of ecosystems that have been wounded by human activity."

Another example of restoration in practice is the stewardship and fish restoration program of the Big Manistee River Watershed by the Little River Band of Ottawa Indians, an Anishinaabe nation in the Great Lakes region. Settler colonialism across the Great Lakes reduced Nmé (lake sturgeon) numbers and threatened Anishinaabe cultural, political, and ecological systems through habitat degradation, over-harvesting, damming, and stocking nonnative species for sport fishing. Marty Holtgren et al. (2014) describe the impetus for restoration in distinctly relational terms. "The fish are ancestors: restoration is about bringing them back to the river and restoring the relationship with Nmé." Since the Sturgeon Program launched in 2001, this plan has become a blueprint for stewardship throughout the Great Lakes, with collaborations from the US Fish & Wildlife Service and Michigan and Wisconsin state agencies. It has also fostered relational repair between Anishinaabek and settler Americans in the Manistee region. "The fish has been able to both heal old wounds and create new, sustainable relationships among people, even in a water-shed where these relationships have been strained by settler colonialism" (Holtgren et al. 2015). Though this perspective is far from universal, Holtgren and coauthors join many other scholars and practitioners who understand the work of ecological restoration as including restoration of human-ecological relationships.[6]

FAKING NATURE AND THE BASELINE PROBLEM

Stuart Allison (2007) realizes that restoration goals can seem quixotic, even arbitrary to outside observers. Do restorationists really take themselves to be returning a place exactly or precisely to a prior arrangement? If so, are we fooling ourselves in thinking we know exactly what once was, believing we can put it all back, as good as new? Ecosystems change, after all. Isn't that at odds with a goal of perfect stasis? Does restoration require privileging one historical moment over all other historical alternatives as *the* moment to be prized? As Marion Hourdequin (2016, 13) says, "restorationists continue to wrestle with the question of whether history is relevant *at all* in a rapidly changing world." With these concerns in mind, Allison responds as follows:

> The hang-up some environmental philosophers express about whether restorations are natural or not, or even whether the natural still exists, misses the point. The connection between humans and the environment cannot be denied. The fact that the relationship is not working well cannot be denied, either. (Allison 2004, 286)

Allison's evocation of human-environmental relationality and its sorry history converges with Margaret Walker's defense of restorative justice. *Given* injustice, how can offenders, victims, and our communities move forward? How can we repair our torn relationships? Putting things this way can inspire some skepticism when our histories of injustice and wrongdoing are such that there *may never have been* healthy relationships between the parties to which to return. To what state of affairs, then, are we trying to return? Walker offers this clarification:

> The terminology of "restoration" is sometimes criticized because it implies a *return* to a condition of relationship that either did not exist or was unacceptable. I propose that we understand "restoration" in all contexts as normative: "restoration" refers to repairs that move relationships in the direction of *becoming morally adequate,* without assuming a morally adequate status quo ante. (Walker 2006b, 384)

Bringing reparative environmental justice to bear on ecological restoration helps to illuminate that relationship, its history, and its repair. What happens when we see ecological restoration not only in terms of recreating particular faunal and floral configurations but as practices building a morally healthier human-ecological relationality?

Conceiving restoration work as repairing the moral conditions of interpersonal, interspecies, and other biotic relationships damaged by ecosystem degradation and destruction can be useful. For example, the emphasis on reparative rather than restitutive justice sidesteps a concern often raised about restoration ecology: namely, can the lost value of degraded ecosystems ever be fully restored? This *restoration thesis* coined by Robert Elliot (1982,

1994) has been subject to much debate in environmental ethics.[7] Even at its best, we ask, can a restored ecosystem ever be as good as it was before? Is there some value that the original ecosystem had which the restored one simply cannot attain? If we see the value of ecological restoration work in terms of rebuilding an ecosystemic configuration that once was, that restitutive approach does invite Elliot's skeptical query. But in applying the relational model of reparative environmental justice, the aim of a restoration project needn't be one of perfect replication. Strictly speaking this might be impossible, and practically speaking few if any working restorationists would ever be so bold as to claim full value had been restored. But more to the point, the question posed by the restoration thesis is not what a relational approach to reparative justice aspires to answer. For restoration projects understood in relational terms, the work is not about restoring lost value but rather taking responsibility for ecological destruction, making amends, and demonstrating renewed trustworthiness.

Related to the restoration thesis is an inauthenticity critique: that is, the claim that ecological restoration *fakes nature* (Elliot 1997).[8] The restorationist goal seems to be to build an artifice that looks indistinguishably natural, after all. When understood and undertaken as part of a process of reparative environmental justice, however, restoration projects are not pretensions to wildness.[9] To feign as if environmental degradation never happened would run contrary to the reparative requirements of apology and acknowledgment of wrongdoing. Furthermore, the emphasis on amends as a process to be performed, rather than an endpoint to be reached, resonates with our recognition of restoration as work, *labor*, an iterative process of construction and destruction, death, and growth. For both ecological restoration and restorative justice, acknowledging our histories and doing the work are crucial.

This raises another familiar issue for ecological restoration, what Alex Lee and coauthors call the "baseline problem" (2014). Here the charge is about arbitrariness: why exactly is *this* specific ecosystem arrangement identified and elevated as our aspiration? The baseline problem informs Markku Oksanen's criticisms of both Paul Taylor and William Jordan on ecological restoration: "Should we compensate for our ancestors' bad deeds and attempt to restore the earth as it was at some point in the past?" (Oksanen 2008, 104). Thinking about ecosystem restoration in terms of restitution invites this line of reasoning, where the state of affairs that is sought predates not just the most proximate case of ecological wrongdoing but our whole sordid history.[10] Reparative environmental justice, by contrast, frames restoration work as normative: restoration in the sense of moving ecological relationships toward becoming more morally adequate. The goal is not an arbitrarily privileged ecosystem nor one that (mythically, romantically) may have existed before all human degradation, but that which is required based on the relevant victims' perspectives and priorities to rebuild trust and repair damaged human-ecological relationships.

Along similar lines, reparative environmental justice provides a response to Katz and others who are worried that ecological restoration shows the same problematic attitudes of imperialistic domination and epistemic arrogance that caused ecological degradation in the first place. Should we not leave well enough alone and let the natural world recover on its own? We cannot dismiss this criticism too quickly. At the very least it calls for epistemic humility on human knowledge of ecosystems and our place within them. But I would argue that a restoration project viewed and enacted as part of a process of reparative environmental justice might be less prone to attitudes of arrogance and supremacy, and that restoration work done to make amends might have a different meaning than one of domination. Those who are actively engaged in restoration work know their projects cannot be controlled in a determinative sense. Like penitent wrongdoers trying to make amends, restorationists have agency in the reparative process, but they cannot and do not effect successful repair on their own. The ecosystem has a life of its own (Throop 2012, 54; Vogel 2015, 111). Plantings give way to unguided dissemination; native species fill spaces evacuated by weeding and prescribed burns; wolves and migratory birds repopulate an area, and so on.

Whether the return of native species and migratory birds to a restored ecosystem should be thought of as interspecies forgiveness or renewed trust is a matter of some dispute. Many of us would regard the notion of forests and swamps as forgiving or trusting as metaphorical at best, while others are less wary of more literal interpretations. Such issues are taken up below. For present purposes, what matters is our recognition that ecological restoration is not entirely under human control even as human actions do contribute to the restorative process.[11] Human beings who actively contribute to ecological restoration cannot force success, any more than offenders can force victims to forgive and forget the harms their destructive behaviors have caused.

PROFESSIONALIZED AND PREPLANNED AMENDS

Accepting responsibility for our actions and recognizing them as wrongful may not be sufficient for moral repair, but these steps are necessary to even begin to set things right. Along these lines, for an action to constitute an attempt at amends, the actor must acknowledge their wrongdoings and offer their action as redress for wrongdoing. Otherwise, Walker says, "reparative actions are charitable, compassionate, or generous, even dutifully so, but they do not 'make amends'" (2006a, 191). Ecological restoration *can* be done with a sense of contrition, and at their best restoration projects enrich the material conditions in which human beings and other creatures live. Yet restoration projects are often organized as relentlessly forward-looking, with no sense of them constituting an acknowledgment of prior wrongdoing (Hettinger 2012, 35). The distance between restoration and acknowledged wrongdoing is perhaps most glaring when restoration projects are planned in advance and even offered as validation for anticipated future environmental destruction. Standard practices of preplanned mitigation would seem to frame restoration as compensation, such that developers need not see themselves as having anything to apologize for: after all, a newly built pond, prairie, or wetland seemingly puts everything right, acre for acre. "Restoration narrowly construed," Mills (1995, 9) warns, "could be just another form of greenwash."[12] How can we take responsibility for and sincerely acknowledge as unjust or otherwise wrong acts of ecological degradation or destruction that we then proceed to carry out?

The issue of unacknowledged wrongdoing could be a problem for reparative environmental justice, but I would argue that it is more of a problem for standardized practices of preplanned restoration as mitigation. Recall Light's (2003a) distinction between *benevolent* and *malicious* restoration. Restoration as preplanned mitigation would seem to be incompatible with genuine contrition and acknowledged wrongdoing, and precisely the sort of restoration projects most susceptible to characterization as malicious. I am inclined to press Light's point, giving greater moral warrant to those restoration projects that can be identified without hypocrisy as practices of moral repair.

Concerns about how ecological restoration might be meaningfully understood as constituting reparative justice extend not only to the work of amends but to who is doing this work. Repair as a deliberate engaged *process* and not merely a victim-centered *outcome* means that wrongdoers seeking to prove their renewed trustworthiness must do the work, literally. They cannot pass off the messy details to others, nor "surrender the role of doing justice to experts, professionals, or 'the state,' which should play facilitating roles" (Walker 2006b, 383). In considering ecological restoration in practice, however, professionalization comes to the fore. Whether employees of state agencies or contractors for corporate, governmental, or residential projects, restoration technicians do most of the physical and intellectual labor. Meanwhile those directly responsible for environmental destruction might be footing the bill, but they are unlikely to be doing any actual labor of restoration. Absent any genuine, actively engaged contrition from those most responsible, where then does this leave those who actually do the burning, weeding, herbiciding, seeding, and planting? How can restoration projects count as making amends if they are not done by those who are responsible for wrongdoing? Should we even expect professional restoration technicians to see their work *as* making amends, or as anything other than ecological improvement? Professionals may feel responsible for *other* destruction or degradation for which they themselves are more directly responsible, to be sure, and yet their work as hired technicians might be completely unrelated to that.

Walker allows that there are indeed situations in which wrongdoers fail to take responsibility, apologize, or do the work needed to rebuild trust. When that happens, "restorative justice invites communities, of varying sizes and descriptions, to participate both as actors in repair and guarantors of repair. . . . When responsible individuals are unrepentant or contemptuous, repair devolves to communities or networks of support within communities" (2006a, 161). Recall also Light's advocacy of the value of public participation in restoration projects: "Public participation does not mean that expertise should be abandoned in restorations; it just means that whenever possible, restorations are better when experts guide voluntary restorationists" (2006, 173).[13] Projects where volunteers do restoration work alongside expert practitioners play a valid role in reparative environmental justice. We can work to make amends as a community even when those who are directly responsible for destruction fail to do so. Reparative justice is as much about our communities' reaffirmation of victims' moral standing as it is about perpetrators' apologies and amends. In this way, we can work toward restoration of the moral conditions for healthy trust relationships in our communities locally, globally, and intergenerationally, without relieving unrepentant perpetrators of their own responsibilities.

REMEDIATION, RITUAL, AND RELATIONAL REPAIR

Recent revaluations of ecological restoration by Basl, Throop, and others offer virtue accounts of the value of restoration. Basl allows that successful restoration as restitution may have what he calls a "reparative requirement," concerned with recreating a specific ecological configuration; but this is distinct from and should not obscure the equally important "remediative requirement" of restitutive restoration, he argues (2010, 12–15). For restoration as remediation, a person must identify the dispositions that led to their environmental wrongdoing, and then in doing the work of restoration thereby correctively cultivate more virtuous dispositions within oneself. As with restoration work itself, the cultivation of virtuous dispositions is a process, not just an outcome, and one that requires not only reflective contemplation but also sustained and deliberate activity. "Particular acts of restoration will be restitutive," Basl explains, "insofar as they are arranged in such a way that they are conducive to character remediation" (2010, 15).[14]

I applaud this more expansive treatment of the moral value of restoration, but I worry about an emphasis on perpetrators' character remediation that stays disconnected to relationality and victims' subjectivity. It is not that Basl sees identifying objects of restitution to be unimportant, only that it is unnecessary for the remediation requirement. For this, we are instructed to locate the wrongdoer's relevant dispositions, but not the relevant wronged parties, nor the relationship in which the wrongdoer and the wronged parties stand. Thus it is reasonable to wonder whether the remediative project will lack definition and direction if it does not attend to the subjectivities of the particular parties wronged and specificities of the relationships at hand. Whose perspective is implicitly or explicitly privileged when identifying the errant dispositions and the appropriate corrective activities for character remediation, if victims' subjectivities and relational specificity have been put aside? Approaching restoration through the metaphor of healing, Throop reminds us that "a good healer exhibits sensitivity to the particularities of the patient ... [especially] when the 'patient' is an ecosystem, interrelated with other systems in the region" (Throop 2012, 56). Here he recommends "humility, self-restraint, sensitivity, and respect for the other" (2012, 48) as among those virtues that are particularly relevant to good restoration.

In analyzing the rhetoric of restoration, Laura Smith finds what she calls "an almost taken-for-granted assumption that 'redemption' implies the expiation of guilt through some form of recompense or payback" (2014, 288). She sees emotionality, for example, in Jordan's account of restoration as "a kind of ritual, even a sacrament, of reentry into nature," whereby we repentant wrongdoers might redeem ourselves.[15] One risk in focusing on ritual and redemption in how we understand and enact restoration projects, however, one which reparative environmental justice can help us avoid, is privileging the experiences and perspectives of perpetrators of ecological destruction. Whether deliberate or not, this priority has real implications for how success in the design and implementation of restoration is framed, including the questions of when, by whom, and whether a particular restoration project should be pursued at all.

Indeed, is the restoration of morally broken relationships even an appropriate goal following ecological damage and wrongdoing? Perhaps some broken relationships should not be repaired. This is somewhat though not wholly akin to Katz's (2012) critique of restoration and Matthews's (1999, 2004) ethos of counter-modernity, which counsel against human domination over nature but not for severing human-nature relations entirely. Abolishing a relationship given ecological destruction might be articulated and defended in at least two ways. First are those cases in which perpetrators are right to acknowledge their wrongdoing and to try to make amends while victims are nevertheless justified in withholding trust and forgiveness. In a second set of cases, relational repair is so unsuitable given atrocious environmental wrongdoings that the contrite perpetrator's very attempt to apologize and make amends is itself inappropriate. Here accountability means the perpetrator must admit their faults and until further notice leave things be, wise enough at least to know that their efforts at resolution would be unwanted and unwelcome. It is not contradictory to reparative environmental justice to grant the possibility that some human-ecological relationships cannot or should not yet be repaired.

PLURALISM AND CHALLENGES OF VICTIM IDENTIFICATION

Theorizing ecological restoration in terms of reparative environmental justice means attending to victims, perpetrators, and ecological-social relationships in their histories of environmental degradation. Margaret Walker's call for repair to center around victims' experiences, views, and expectations serves as a safeguard against their marginalization in the planning, implementation, and overall continuity of restorative practices. Successful restorations are not solitary human impositions of artifice: restored savannas, lakes, and wetlands are importantly unlike playfields and playgrounds in that sooner or later, an ecosystem must take on a life of its own. Ecological restorationists cannot force ecological flourishing, as contrite wrongdoers cannot force victims' trust or forgiveness. And maintaining focus on the victims of environmental degradation as the beneficiaries of ecological restoration provides much needed clarification and direction. There is a tendency to assess the success of restoration projects in diffuse terms: general public utility of restored landscapes might be cited, for example, with no attention to the particular relationships among various parties or particular histories of degraded and restored places (Throop 2012, 56). Rather than generic public good or perfectly pretend wilderness, reparative environmental justice is organized around the particular needs, values, and experiences of those wronged by ecological degradation.

Of course, this prioritization only works if ecological degradation and destruction can be said to have victims. Oksanen's second criticism of restoration rejects this: "If we consider nature as a victim of our wrongful actions, we lack the victim," he writes. "Nature lacks victimhood, that is, the essential psychological and complex qualities that characterize the life and the experiences of human victims and the nature of human communality" (2008, 105). While he is more optimistic than Oksanen on the merits of restoration, Basl

(2010, 11) recommends putting aside the thorny issue of victim identification as irrelevant to the remediative potential of restoration.

While victim identity is a challenge, I believe a good dose of environmental-ethical pluralism can help here. Different environmental philosophies assign different moral standing to individual humans, natural entities, social groups, and ecological systems. And yet each of these assesses environmental destruction as harmful: to ecosystems directly, to individual creatures negatively impacted, to human beings directly or indirectly hurt by ecosystem degradation. The core idea of restoration as part of repairing the conditions of our moral relationship holds, but explicated differently for different environmental ethics in terms of the victims and the relevant moral relationships to be identified.[16]

Those who are sympathetic to more traditional anthropocentric ethics and are understandably wary to describe human relationships with other animals and ecosystems in terms of forgiveness or trust might nonetheless find a meaningful version of reparative environmental justice befitting of their ethics. An anthropocentric ethic that is alive to environmental racism and environmental injustice, for instance, can make sense of reparative environmental justice among human beings. Environmental-justice activists rightly identify indigenous groups wronged by degradation and destruction of their traditional lands, waters, and traditions (Gilio-Whitaker 2019). Victims of environmental degradation also include future people. Ecological restoration can be understood as making amends to past, present, and future people; as environmental-justice campaigns often extend across racial, class, and national inequities, we might recognize them as extending across international and intergenerational inequities too (Buxton 2019). Meanwhile animal welfarists and animal-rights theorists who recognize individual animals' moral status may take up versions of reparative environmental justice that recognize human and nonhuman animals as identifiable victims of environmental harm and interspecies conflicts. On a non-anthropocentric ethic, the possibilities of interspecies amends, trust, and forgiveness might not seem so implausible. For their part, environmental holists may adopt suitably holistic versions of reparative environmental justice, such that restored habitats are themselves viewed as party to restored ecological-ethical relationships, and practices of ecological repair are directed toward not only individual human or nonhuman animals but also the ecosystems of which these individuals and groups are part. Those who adopt an eco-theological ethic may also adopt a theological version of environmental repair, one identifying God or gods as those to whom penitence is owed and from whom forgiveness is sought, toward renewed fidelity in human-divine relationships.

The pluralistic suggestion is that those differently motivated can work capably in coalition on ecological-restoration projects, finding some overlapping consensus in approaching such projects as ways of making amends understood differently given contributors' differing ethics. As with any coalition, this overlapping consensus can be fragile and different versions of environmental repair befitting our varied environmental philosophies will not always give the same ecological or ethical recommendations. Pluralist environmental repair and victim-centeredness might prove more or less workable for some projects or coalitions than others. Complications might well persist. Walker herself advocates an explicitly human-centered feminist approach to morality, which animal-rights ethics, environment holism, and eco-theological ethics will all depart from. Thus it remains an open question whether this constitutes too great a departure. These warnings notwithstanding, a contingent, coalitional, open-ended approach is a promising place from which to start.

My hope is that readers may see the value of this model of reparative environmental justice put into practice when we return to the case of ecological restoration in the Chicago Wilderness in chapter 7. Yet I would be remiss if I ended this chapter without acknowledging and engaging Katz's recent criticisms of my reparative account. "Ideas of restorative justice and moral repair are appropriate to address human injustice and wrongdoing," Katz allows. "But these concepts are vacuous and lose their meaning when addressing the ethics of human activities regarding the natural world because of the essential character of the restoration process: the replacement or substitution of new entities for pre-existing entities in an attempt to reverse the irreversible" (2018, 17). Here he does not mean that ecological restoration necessarily replaces or substitutes red oaks for white oaks, prairies for wetlands, or one kind of salmon for another. For Katz what is really essential is that restoration creates *artifacts* in place of nature. Restored ecosystems are deceptive or delusional (or perhaps both) in the sense that restorationists are trying to pass off something built as though it were natural. What makes restored ecosystems artificial is not that the trees are plastic, the grass is genetically modified, or the waterfowl nesting along a restored lake are really wooden ducks. What makes them artificial, Katz says, is that they come from the same sort of human arrogance and domination that damaged the ecosystems in the first place. We cannot reverse the irreversible, Katz concludes; the best we can do is to let things be and let the natural world heal itself.

The natural/artificial distinction at the center of Katz's critique of ecological restoration has been interrogated by critics and defended and refined by Katz himself.[17] In its later formulations, Katz specifies that some human activities can indeed be natural, while restoration not only can be but essentially *is* artificial. Negatively I would echo Donna Ladkin's (2005, 209) assessment that restoration need not imply domination; positively I agree with Robin Kimmerer that "restoration can be viewed as an act of reciprocity, where humans exercise their care-giving responsibility for ecosystems" (2011, 257).[18] As I have argued in preceding sections, part of what is distinctive and clarifying about a relational-reparative approach to ecological restoration is that it reframes our assumptions about what needs to be repaired, and why. As Kimmerer puts it, "It is not the land that is broken, but our relationship with it" (2011, 272). If we take restoration to be the work of making amends, rather than a futile attempt to put everything back as it was, then the goal is not reversing the irreversible but repairing damaged human-ecological relationality. So understood, the historical record for a specific place is relevant not as something to be forged or faked but as defeasible evidence of what sorts of material configurations promote the flourishing of various plants, animals, and people who live in the ecosystem. As Zentner (1992, 114) put it in an earlier response to Katz, "Origin is less important in judging the authenticity of a restored landscape than is its behavior or performance." Whether a restored prairie, for example, is properly understood as natural or artificial might matter if the key question is whether the restored prairie is identical or equivalent to the prairie that existed prior to human wrongdoing. But if our approach is one of reparative environmental justice, not compensatory or restitutive justice, then what matters are the ways in which the process of prairie restoration might help or hinder relational repair for the human and nonhuman members of the biotic community.

Katz's objections to ecological restoration as moral repair go beyond the natural / artificial distinction. He further argues that ecological restoration as moral repair has a major problem in "the absence of a 'partner' or 'victim' on the other side of the equation" (2018, 20). Katz does acknowledge my pluralist proposal, that different environmental ethics may identify victims of ecosystem degradation and destruction to whom our apologies and amends would be directed, but he is unpersuaded that this approach can give a workable solution to the challenge of victim identification. Any victims deserving reparations are gone, he says; moral repair is impossible. "How can we use the ideas of repair, trust, and forgiveness in a relationship with a newly restored ecosystem when the entities receiving the repair and offering the forgiveness are different from the original entities harmed?" (2018, 22)

I certainly agree that some, perhaps even many of our relationships damaged or destroyed by ecological degradation may be beyond moral repair through ecological restoration. This is true for multiple reasons, including but not limited to the fact that ecological restoration does not help nonhuman victims of environmental wrongdoing that no longer exist. Yet Katz's inference to the impossibility of repair is overstated. Some victims no longer exist, but of course others do. On an eco-theological ethic, the prospects for renewed fidelity in human-divine relationality are not undercut by ontological differences between a restored ecosystem and its earlier form. Whether we take an anthropocentric or non-anthropocentric approach to environmental justice, victims of ecological degradation include (though are not necessarily limited to) human beings whose lives, cultures, knowledges, and other practices have been negatively impacted, and those who seek to apologize for the ecological degradation and human suffering they caused and to make amends through restoration work may direct their reparative efforts toward these human victims. Plants and animals negatively impacted by ecological degradation may also return and flourish again in—and in so doing, also contribute to the health and healing of—restored ecosystems. Seeds lying dormant underground sprout after a series of prescribed burns; endangered dragonfly populations just barely holding on thrive amid newly restored native grasses; migrating cranes once again stop safely at wetlands and prairies rebuilt and recovered from farmland and munitions-testing grounds.

IN DEFENSE OF NONIDEAL ECOLOGICAL RELATIONSHIPS

"We can and ought to help nature heal from this assault and the restoration movement is a praiseworthy acknowledgement of that power and responsibility. But restoration is a short-term and fundamentally regrettable way of relating to nature. While much can be learned from the movement to restore nature . . . restoration does not provide a paradigm for the ideal human relationship with nature," Hettinger (2012, 41) argues. "An ideal community would not need such institutions" (2012, 39).

Yet environmental ethics for ideal communities and ideal relationships is inherently limited. Ideally, we would never do ourselves, each other, or nature wrong. But we have, we still do, and we will again, and so we need ethical guidance that is applicable to life after environmental wrongdoing. For this, we need nonideal theorizing, and my suggestion has been that a model of reparative environmental justice might be especially useful for making sense of restoration work—not as ecological forgery, or attempted compensation, but as the work of relational repair. For many ecological-restoration projects, framing them in terms of reparative justice can be ethically and ecologically clarifying. A relational-reparative approach to ecological restoration requires us to acknowledge our acts of environmental damage and degradation as wrong, to do the work of making amends through victim-guided reparative practices, and when warranted, to renew trust and extend forgiveness toward healthier human-ecological relationships. This approach will not align with all restoration projects as actually practiced, but this is a good thing, as it opens space for targeted criticism and improvement of misdirected, opportunistic, or otherwise ecologically and ethically inadequate projects.

NOTES

1. See also Cairns and Jackson (1996), Palmer et al. (1997), and Hobbs (2005).
2. See also Elliot (1982) and Elliot (1997).
3. See also Light (2000a) and Hertog and Turnhout (2018).
4. See also Throop (2012), Tanasecu (2017), and McLaren (2018).
5. See Schroeder (2000), Light (2000b), Higgs (2003), and Light (2006, 2008).
6. See Jordan (1992), Cairns (1995), Meekison and Higgs (1998), Strohmeier (2000), Hourdequin and Wong (2005), Egan et al. (2011), Kimmerer (2011), and Congdon (2014).
7. See also Gunn (1991) and Light (2000a).
8. See also Rassler (1994), Perry (1994), Jordan (2003), Light (2003), Hilderbrand (2005), Egan (2006), Parrillo (2008), and Hourdequin and Havlick (2013).
9. "Perhaps the most frustrating aspect of Katz's arguments," writes Steve Rassler (1994, 117), "is his complaint that restorationists are trying to 'pass off' restored habitats as being equivalent to undisturbed habitats of the same type. I am unaware of any foundation to this claim." See also Vogel (2015) on transparent restoration.
10. On restoration as restitution, compensation, or "wiping the slate clean," see Cronon (1996) and Wiegleb et al. (2013).
11. On ecological restoration as dynamic and uncontrolled, see Packard (1993), Clewall (2000), Plumwood (2002), and Hilderbrand (2005).
12. On preplanned mitigation and corporate appropriation of ecological restoration, see Light (1994), Perry (1994), Berry (1998), Windhager (1998), Cairns (2003), and Hilderbrand (2005).
13. See also Miles (1998).
14. On the redemptive power of restoration, see also Ross (1994) and Higgs (2011).
15. See Jordan (1986, 25), Smith (2014, 289), and also Hettinger (2012, 39).
16. On pluralism and inclusion in restoration, see Meekison and Higgs (1998) and Cabin (2007).
17. For criticisms of the natural/artificial distinction, see Zentner (1992), Rassler (1994), Light (2000a), Ladkin (2005), and Congdon (2014); for defenses and clarifications, see Katz (1993, 2002b, 2012) and Siipi (2003, 2008).
18. See also Palamar (2006) and Spelman (2008).

Chapter 4

Animal Ethics and Contexts of Interspecies Repair

Though many of us have constructed our lives (or have had them constructed for us) such that it is easy to forget or at least ignore, human lives are entangled with other animals in many ways, invaluable and trivial and everything in between. Some human-nonhuman relationships would arguably still exist in some form or another under an ideal conception of animal ethics, while others have an inescapably nonideal character: these relationships exist as they are *because* things have gone wrong. Of these distinctively nonideal sites of interspecies relationality, some are sites of further harm and suffering, some are sites of interspecies repair, and some may be both, even at once.

ANIMAL RIGHTS, LIBERATION, AND AMELIORATION

Moral relationships between human and nonhuman animals have long been a central topic in environmental ethics. Here attention is paid primarily to queries on nonhuman animals' moral standing or lack thereof, the extent of our obligations to them, and the rightness or wrongness of how we treat and regard them. Less attention is paid to what should happen if—and inevitably when—we fail to give other animals their rightful moral consideration, fail in our obligations to them, mistreat them, or otherwise do them wrong. This relative silence is striking given that proponents of otherwise quite divergent moral philosophies readily agree that this is egregiously, rampantly the case today, and has been for a long time.

Consider the familiar idea of animal rights, starting with its most literal manifestation (as not all animal-rights advocates broadly construed subscribe to explicitly rights-based animal ethics). We do animals wrong when we violate their rights, Tom Regan argues, and as experiencing "subjects of a life," individual animals (human or otherwise) have a right not to be harmed:

> Subjects-of-a-life not only are in the world, they are aware of it and aware, too, of what transpires "on the inside," in the lives that goes on behind their eyes. As such, subjects-of-a-life are something more than animate matter, something different from plants that live and die; subjects-of-a-life are the experiencing center of *their* lives, individuals who have lives that fare experientially better or worse for themselves, logically independently of whether they are valued by others. (Regan 2003, 93)

Regan recognizes that there are situations in which one individual's rights conflict with another, not merely that overriding someone's rights will promote overall utility. In such situations, there is no ideal solution, no course of action that avoids violating someone's or something's rights. What morality requires of us in such cases, Regan explains, is to minimize rights violations. "*Precisely because* each is to count for one, and no one for more than one, we cannot count choosing to override the rights of B, C, and D as neither better nor worse than choosing to override A's right alone" (1983, 305).[1] Three is greater than one, after all, assuming comparable harms involved. Harms are not always comparable, of course—"Other things being equal, M's death is a greater harm than N's migraine" (1983, 311)—in which case, Regan says, the severity of rights violations is what matters.

Putting Regan's mini-ride and worst-off principles into practice can be difficult, but for the sake of argument, let's say his diagnosis is basically correct. What then? In such cases, it's not merely that someone loses and others win, someone is frustrated, or someone's interests are sacrificed for the greater good. By hypothesis, someone's *rights* have been violated. Even if this was the morally best course of action available, its aftermath calls for some moral reckoning; even if we made the right choice, this choice can generate undesired collateral damage, what Walker calls "residues and carryovers" (1995, 145) for those involved. "Recognizing that damage, seeing it for what it truly is, and confronting what we might do about it," Emmerman argues, "is a crucial part of navigating inter-animal conflicts" (2014b, 162). The need for interspecies amelioration is not limited to cases in which individual rights directly conflict. As animal-rights theorists and other ethicists have argued in detail, our conventional uses of nonhuman animals from agriculture to experimentation to entertainment are built on rampant and routinized violations of their rights. This is the core of the abolitionist argument advanced by Regan, Francione, and others, that even if conventional practices could be made less cruel or painful, they can never be made morally acceptable, predicated as they are on regarding and treating other animals "merely as means to human ends, as resources for us—indeed, as renewable resources" (Regan 2003, 97).[2]

But these practices and their accompanying rights violations continue, of course; more to the point, even if abolition were to succeed tomorrow, they have already been done today, yesterday, and years before. It's curious then that there has been little if any discussion of the aftermath of animal-rights violations from an animal-rights perspective. Perhaps this silence is predicated on the presumption that there is nothing to be done in terms of remediation. We can certainly pursue amelioration in terms of relative improvement or iterative adjustment, to learn from the sobering example of animal-rights violations and do better going forward. But is corrective justice even possible, one might ask with a mix of skepticism and despair? Once an animal has been killed and eaten, can any sort of meaningful restitution even be attempted?

To this we can offer multiple responses broadly consistent with a rights-theoretic approach. First, killing aside, there are also nonlethal practices that violate individual animals' rights, in which case perpetrators seeking to rectify their wrongs can face their victims. Second, as Julia Gibson (2019) reminds us, when it comes to anthropogenic climate change, we should not neglect the importance of justice for the dead and dying. And third, the analogy with human wrongdoing reminds us that even when perpetrators cannot make amends to those most directly victimized by their actions, those indirectly affected may deserve recognition, acknowledgment, and apologies as well or instead. For these reasons, those committed to an animal-rights theoretic perspective should not put aside the need for interspecies corrective justice.

Donaldson and Kymlicka argue that animal-rights theory (ART) tells at most half the story, focused as Regan, Francione, and others are on nonhuman animals' negative rights (not to be killed, not to be property, not to be treated like a thing, etc.) and humans' negative duties not to violate such rights:

> By contrast, ART has said little about what *positive* obligations we may owe to animals—such as an obligation to respect animals' habitat, or obligations to design our buildings, roads, and neighborhoods in a way that takes into account animals' needs, or obligations to rescue animals who are unintentionally harmed by human activities, or obligations to care for those animals who have become dependent upon us. (Donaldson and Kymlicka 2011, 6)

Duties of assistance, rehabilitation, restitution, and reparation in the aftermath of rights violations are all candidates for nonideal positive duties that animal-rights theorists should not ignore.

Perhaps the focus on negative rights and duties is strategic: getting people to accept animals' rights not to be mistreated is difficult enough without the further claim that they have rights to be assisted or otherwise treated in a positive way (Dunayer 2004, 119).[3] As Sapontzis puts it, "How these most unfortunate animals are to be treated after they have been released from the current travail is a question for a much better world than ours" (1987, 83). But I am not sure that these two projects can be wholly divorced, unless animal ethics is confined to ideal theorizing, which runs contrary to the unifying endeavor of animal advocacy. It is not as though Regan, Francione, and other animal-rights theorists argue that it *would* violate individual animals' rights *were* we to eat them, experiment on them, or otherwise treat them as things, but that we *are* (and *have been*) doing these things, and it needs to stop. Even putting aside other positive rights and duties, our duty to *stop* animal-rights violations brings along positive duties of amelioration that would not accompany a duty not to commit animal-rights violations in the first place. The need for strategic advocacy might justify sometimes emphasizing the need to stop existing animal exploitation, but those committed to putting ART into practice ought to recognize the aftermath of animal-rights violations as itself a context for moral evaluation. Against Sapontzis, understanding and enacting our obligations to nonhuman animals we have wronged cannot be postponed until a better world than ours is built, for at least two reasons. For one, doing right by those we wronged is part of how we go about building a better world. Furthermore, the issue of how to treat abused and exploited nonhuman animals cannot wait until after abolition, if only because there *already are* abused and exploited animals who need care and consideration, and caretakers working to provide it, day in and day out.[4]

Donaldson and Kymlicka hypothesize that the focus on negative animal rights is not just an oversight or strategic choice but an indication of something deeper about how Regan, Francione, and other animal-rights theorists see human-animal relationships. Abolitionism would arguably put an end to animal husbandry, pet ownership, hunting, vivisection, and interspecies interactions more generally that are abusive, objectifying, exploitative, and otherwise wrong. Would this then put an end to human-animal relationships altogether? Here I will echo Donaldson and Kymlicka (2011, 8), that human-animal relationality is not limited to exploitation and objectification, and furthermore that as ecological beings we cannot help but live in relation with other animals, including not only domesticated but also wild and liminal populations. Lori Gruen puts this point especially well:

> We can't make sense of living without others, and that includes other animals. We are entangled in complex relationships and rather than trying to accomplish the impossible by pretending we can disentangle, we would do better to think about how to be more perceptive and more responsive to the deeply entangled relationships we are in. (Gruen 2015, 63)[5]

In a world without cattle farming, bull fighting, canine police units, or lab rats, our lives would not be entangled with other animals as we are now, but we would be entangled nonetheless. And more to the point, even if abolitionism were to successfully end exploitative or otherwise abusive human interactions with other animals, the moral residue of these prior exploitative or otherwise abusive interactions demands some kind of moral reckoning. We have special positive duties to animals whose rights we have violated *because* we have violated their rights. Even if animal-rights theorists prefer to envision other animals living in minimal relationality to human beings, the long history of interspecies wrongdoing to which animal-rights theorists rightly draw our attention requires them to attend to nonideal rights and duties of interspecies corrective justice as well.

I do not mean to give the impression that animal-rights theorists have said absolutely nothing about the moral aftermath of animal-rights violations. In his preface to the second edition of *The Case for Animal Rights*, Regan (2004a, xl) considers the possibility of compensatory justice for individual members of species with numbers in decline owing to human wrongdoing. Francione (2000, 194) briefly discusses the prospect of criminal punishment for those who treat animals as resources, though mainly to dispel the assumption that criminal prosecution for violating animal rights would be the same as that for violating human rights. The pros and cons of retributive and restitutive approaches to interspecies corrective justice warrant a more detailed discussion than either of these brief treatments provide, but it is still encouraging to see some attention paid to the amelioration of human-animal wrongdoing from a rights-based perspective.

Let's turn from animal rights to animal liberation, whether articulated in explicitly utilitarian or other consequentialist terms.[6] While generally more reformist than abolitionist, liberationists like Peter Singer do agree with many rights-based criticisms of conventional uses of nonhuman animals, though on different grounds. Their problem with animal experimentation, for example, is not the violation of mouse, rabbit, or chimpanzee rights not to be killed or treated like things but the pain and suffering they endure, not to mention the various (physical, cognitive, emotional) positive experiences of which they are deprived.[7] Similarly their problem with eating animals is not so much that we consider them ours to be eaten but rather the suffering to which we subject them in the particular ways that we confine, raise, and kill them.[8]

One might wonder whether utilitarian and welfarist views of animal ethics have any room for corrective justice. Shouldn't animal liberation just focus on interspecies amelioration as relative improvement and iterative adjustment, doing better going forward in promoting animal welfare in our individual and collective actions? If so, then isn't it basically irrelevant whether animals are or are not victims of past wrongdoing? I agree that interspecies amelioration may look quite different for utilitarian than rights-based animal ethics. But is there really no place for corrective justice in animal liberation? Here I find worth revisiting Donaldson and Kymlicka's point about the curious asymmetry between our negative and positive duties to animals: in theory, utilitarian animal ethics should be equally concerned with what we should *do* to animals and what *not* to do them, though in practice this hasn't been the case.[9] If we should not cause unnecessary pain and suffering, what if anything should we do? Whatever actions of assistance to other animals may or may not be required by utilitarianism in ideal circumstances, once human-animal wrongdoing is recognized as not just possible but an extant reality, refraining from harmful action cannot be the whole story.

I have argued that despite their relative neglect of the issue, those who approach animal ethics from rights-based and consequentialist perspectives have reasons to care about what we owe and how we ought to treat other animals *given that* we have significantly, repeatedly wronged them. Meanwhile there are other approaches to animal ethics, including ecofeminist, care-based, and relational analyses, which by comparison have devoted more attention to the moral aftermath of human-animal wrongs. Feminist ethicists such as Carole Adams, Karen Warren, and Lori Gruen examine the interlocking oppression and domination of women, nonhuman animals, and nature generally, with the goal not only to diagnose but to dismantle man-made oppressive structures.[10] One distinct feature of feminist care ethics is its emphasis on our actually existing relationships and their attendant moral responsibilities. As Annette Baier (1986) reminds us, only some of our relationships are voluntary, and yet those relationships into which we are born or are thrust upon us demand care and consideration too.[11] This does not mean we are obligated to maintain all our relationships as they are: as Gruen (2015, 64) puts it, "Inevitability does not entail immutability. Being necessarily in relationship doesn't mean we are completely determined by them or that they are fixed in ways that can't be changed." But it does mean we cannot avoid interspecies relationality simply by refusing to enter into relationships. The question is not whether to be in relation to other animals, pattrice jones reminds us, but rather how we want these relations to be. "Are you content for those relationships to be characterized by domination, violence, and disregard?" jones asks. "Are you comfortable with the habits of belief and behavior fostered by your own speciesism?"[12]

TWO ACCOUNTS OF INTERSPECIES REPAIR

Among animal ethicists today, Karen Emmerman and Clare Palmer are notable for their efforts to adapt human-centered accounts of moral repair and reparative justice for interspecies contexts. For her part, Emmerman offers an ecofeminist account of the distinctive obligations we have to other animals following moral dilemmas and other conflicts. Recall Paul Taylor's compensatory approach to balancing human exploitation of the rest of the biotic community, where restitutive justice means "bringing about an amount of good that is comparable (as far as can be reasonably estimated) to the amount of evil to be compensated for" (1986, 305). Against Taylor, Emmerman (2014a, 220) warns that treating an amount of evil as something that can be compensated invites a dangerous sort of moral complacency, the delusion that perhaps there was no wrongdoing after all. We might worry that compensation frames our moral response to wrongdoing the wrong way and sets the wrong benchmarks for moral evaluation, especially when it comes to interspecies relationality.

Let us consider animal sanctuaries as an ameliorative response to interspecies wrongdoing.[13] Compared to a zoo or aquarium, the mission of an animal sanctuary is not so fragmented. As Dale Jamieson (2002) reminds us, the value of zoos and aquariums can be difficult to pin down in part because these facilities are commonly expected to fulfill a variety of functions, including education, entertainment, animal welfare, and endangered-species preservation.[14] Not all zoos are equally successful in these varied functions, nor do they all pursue all of them in their organizing missions. It would be odd indeed to justify the existence of a small-town zoo with no endangered species by appeal to endangered-species recovery programs at the San Diego Zoo, for example, no matter how successful these programs might be. By contrast, while some animal sanctuaries permit public visitation and others do not, they all share an overarching organizing principle of animal care.

On the surface Taylor's compensatory model of interspecies restitution might seem a decent fit for animal sanctuaries; indeed, some proponents praise and celebrate sanctuaries in just such terms.[15] But the necessity of captivity complicates the picture of animal sanctuaries as sites of restitution. While unquestionably better living conditions than in most animal research facilities, even well-run sanctuaries involve captivity, confinement, problems of boredom, and significant limitations or outright prohibition of many natural activities. Generally speaking limitations on animal freedom and welfare owe not to caretaker apathy nor antipathy but practical necessities. "Though sanctuaries do their best to mitigate these aspects of their animals' lives," Emmerman (2014a, 225) concludes, "it is unavoidably true that we are unable to alleviate all of the harms of captivity."

We may reasonably worry that when sanctuaries are celebrated as sites of restitution, where wronged animals are made whole again, policymakers will see them as adequate counterbalance to continuing and anticipated animal suffering and exploitation elsewhere. By contrast, animal sanctuaries understood in terms of interspecies moral repair serve neither to counterbalance nor excuse animal exploitation elsewhere, nor to hide the reality of sanctuary living conditions. One chimpanzee advocacy organization puts the point this way:

> Even at the best sanctuaries, once removed from the wild or bred in captivity, no chimpanzee can ever be truly free again. We can never give back what was taken from them: the right to be free and live autonomously. This reality makes the imperative to do right by them, even within the boundaries we have imposed on them, all the more urgent and mandatory. (Project R&R 2019)

The aim of reparative justice is not compensation but to rebuild renewed healthy relationality between human and nonhuman animals. Unlike with compensation, we recognize the work of interspecies repair as dynamic and ongoing just as interspecies relationships themselves are dynamic and ongoing.

Sanctuaries are an important context of interspecies moral conflict and amelioration, but not exclusively so. Emmerman (2014b) also invites us to consider a mother who, despite her strong commitment to nonhuman animals' moral standing and the great lengths she takes to find other solutions, must use animal-based products in the urgent care of her infant son. The main point of concern here is not that she made the wrong choice, nor that she should have acted otherwise, but what to do with the moral residue that remains *even if* her choice was actually warranted. Talk of compensation rings hollow: restitution to the actual animals harmed often proves impossible and anyway this seems to mischaracterize the moral remainder under consideration here. Rather than try to "pay back" these animals, Emmerman argues, the mother seeking to make reparations can strive to demonstrate her recognition of their moral significance, deep gratitude, efforts to make amends, and hope in the possibility of renewed interspecies relationality.

Sanctuaries and rehabilitative facilities work with a variety of animals, but generally speaking they fall into two categories: agricultural and other domesticated animals (chickens, cows, pigs, horses) and nondomesticated animals whose mistreatment by humans has rendered them unable to live on their own (chimpanzees in laboratories, former circus animals, once wild coyotes now imprinted on humans). What about wild animals: what duties of assistance do we have to them? Palmer says that most people agree that we have duties of assistance to domesticated animals we do not have to wild ones. She calls this the *laissez faire intuition*, and in *Animal Ethics in Context* and several subsequent articles, Palmer attempts to explain why this might be so.[16]

Her argument comes in two main parts. First, while we have prima facie negative duties not to harm nonhuman animals, we also have positive duties of assistance to those animals with whom we stand in some kind of relationship, including "human/animal entanglements, histories, and shared institutional frameworks where humans are, or have been, either responsible for harms to animals or for the generation of particular vulnerabilities to animals" (2010, 88–89).[17] We have duties to domesticated animals not merely if and when we actively harm them, but because we are responsible for their domestication and attendant vulnerabilities. Our confinement prevents them from doing many essential things for themselves (finding food, mating, etc.) but taming also has changed their natures, including their very capacities for self-sufficiency. Consider the difference between captive wild animals, who *could* care for themselves in their native habitats but are dependent on human assistance while confined, and the *internal* dependencies created in domesticated animals (Palmer 2010, 91). I think it is worth complicating Palmer's distinction a bit here, recognizing that entanglements with human beings can affect a wild animal's nature and thereby produce vulnerability even without domestication. So when children feed a young turkey vulture, for example, the wild bird is not tamed but it might be imprinted, and detrimentally so, now no longer able to live safely outside captivity. Chimpanzees and other animals from captive breeding programs may not have centuries of domestication like cats and dogs, but having been raised from birth in human facilities, it's unlikely they could fend for themselves in their native habitats either.[18]

Second, Palmer proposes that we can extend the lessons of human-centered reparative justice to our duties to other animals—specifically, the lesson that past harms create special obligations. "Since I have argued that animals can be wrongfully harmed, a question about reparation—or at least some backward-looking special obligation—is not ruled out in principle" (2010, 99).[19] For example, consider coyotes displaced from their habitats by human development projects:

> Human activity has had the effect of making the coyotes more vulnerable (by exposing them to new threats such as road hazard and hunting) while at the same time compromising their ability to provide for themselves by building on their habitat, reducing the numbers of their prey animals, destroying denning areas, etc. In this sense, the human effect on coyotes (in intensifying vulnerability and reducing self-sufficiency) seriously sets back their interests, in an ongoing way. If we take their interests with moral seriousness, these harms should generate some backward-looking special obligations to assist. (Palmer 2010, 102–03)

Palmer grants that there are differences between this case and human-centered reparative justice, to be sure. For one thing, since coyotes cannot know that reparation *is* reparation, "anything like an apology or a memorialization would be wasted on them" (2010, 104). And since nonhuman victims take no satisfaction in seeing contrite perpetrators make amends, there is no reason why reparations need to be made by perpetrators specifically, as long as they are made. Sometimes a perpetrator-focused approach to animal reparations is called for, Palmer allows, and other times we have reason to take a beneficiary-focused approach instead.

Those who agree with Palmer's account of our duties of interspecies assistance and reparation might disagree about whether the laissez faire intuition follows from it. This is Trevor Hedberg's criticism, that the premises of Palmer's argument do not entail her intended conclusion about our different duties to domesticated and wild animals. As an alternative to the laissez faire intuition, Hedberg (2016, 428) proposes a *gradient view* according to which we have a weak presumptive duty to assist wild animals.[20] Palmer's reasoning here is that "there is no analogy to current or historical unfairness (or, indeed, fairness or justice) about the states in which wild animals find themselves. Inasmuch as they live without human contact, they are outside the realm of justice altogether" (2010, 89). One worry is that anthropogenic climate change and other wide-sweeping environmental damage and degradation means that few if any animals will count as fully wild on this definition. Palmer considers the implications of climate change for her view, and argues that even if wild animals are "contacted" this does not necessarily mean we have duties of assistance to all of them, since some will actually benefit from global climate change even as others will be harmed by it (2010, 142).[21]

ON INTERSPECIES TRUST AND FORGIVENESS

For her part Margaret Walker takes an explicitly human-centered view of morality: as she puts it, "Morality is fundamentally interpersonal: it arises out of and is reproduced or modified in what goes on between or among people" (1998, 10). While this stance could possibly accommodate nonhuman persons, elsewhere her human-centered approach is clear:

> Because I believe morality consists in interpersonal acknowledgment and constraint, from which people learn that they are responsible for things and to others, I cannot think of it as something that could obtain outside human relations and humans' experiences of them. Morality arises and goes on between people, recruiting human capacities for self-awareness and awareness of others' awareness; for feeling and learning to feel particular things in response to what one is aware of; for expressing judgment and feeling in the responses appropriate to them. . . . But I do not give up the right to talk about moral reality, because I

think morality is a strikingly real dimension of every human group's social life. (Walker 1998, 5)

Non-anthropocentric ethics would appear to be at odds with this view of morality. In extending and revising reparative justice into interspecies contexts, we are free to adopt different views on the place of nonhuman animals in moral relationships, of course. But in this case, is it too great a departure?

Consider interspecies trust and forgiveness, which are challenges not only for anthropocentric ethics but also for those approaches to animal ethics that affirm nonhuman animals' moral standing yet balk at ascribing capacities for trust and forgiveness to them. We can imagine a variation on Jeremy Bentham's famous query often cited in discussions of animal ethics, rephrased this way—the question is not *Can they trust?* nor *Can they forgive?* But rather *Can they suffer?* Sentience overlaps with but is not equivalent to capacities for trust and forgiveness. How then can animal ethics make sense of reparative justice for those nonhuman victims who meet the sentience but not forgiveness or trust criteria?

Here I think we are left with at least two (not mutually exclusive) possibilities. The first is to limit ourselves to interspecies reparative justice where the wronged nonhuman parties to whom apologies and amends are owed are able to trust and forgive us in a meaningful sense, including chimpanzees, dolphins, and other highly intelligent subjects of experimentation, entertainment, or other forms of captivity and confinement. The second option is to jettison the requirements of forgiveness and trust in cases of interspecies reparative justice. One disadvantage of this second option is that it would seem to marginalize victim subjectivity, which is a significant source of normative force and direction for reparative environmental justice as I have described it. For this reason I suggest a hybrid view incorporating both possibilities, such that interspecies reparative justice calls for renewed trust and possible forgiveness but just in case the particular interspecies relationships and nonhuman victims' subjectivities allow for trust and forgiveness.

Even this hybrid proposal needs some further clarification, however. The fact that an animal has a capacity for trust does not mean that its relationship with human beings should be trustful. Even those of us who urge recognition of human-animal relationality must emphasize that trust is often not what such relationships need. The examples of human wrongdoing against liminal and wild animals remind us that, important as they are, trust and forgiveness are not essential for all relational repair, because not all of our interspecies relationships to be repaired should be based in trustful dependency. Renewed trust and forgiveness are important when victims done wrong are capable of trust and forgiveness and when the nature of the moral relationships involved calls for them as conditions for morally healthy interspecies entanglement.

Relational repair can still be an appropriate ameliorative goal when things are otherwise, and the specific subjectivities of nonhuman victims can still be prioritized in a reparative process. In such cases, instead of asking what animals need to rebuild trust, we ask the more general question of which renewed trust and trustfulness is a specific variant: what do they need in order to restore (or perhaps create) more adequate moral relationality among human perpetrators, nonhuman victims, and the rest of the biotic community of which we are part?

ANIMAL CARE, APOLOGIES, AND AMENDS

Whether animal sanctuaries are properly understood as reparative practices concerns not only the work of amends but also who does this work. Moral repair is a deliberate and engaged process, not just an outcome, which means that wrongdoers seeking renewed trust and forgiveness must do the work. They cannot just pass off the messy bits. Sometimes the work of interspecies repair is done by the perpetrators of precipitating acts of animal suffering. Consider pets, farm animals, or service animals who have suffered harm, and the humans responsible for that harm working to acknowledge their responsibility and rebuild trust with these animals. But for animal sanctuaries, as they tend to be organized and implemented today, professionalization is the norm, and rightly so, with skilled, experienced caretakers doing the day-in/day-out work. Those who are directly responsible for interspecies wrongdoing *might* be footing the bill, though not always, and either way is not likely to do actual rehabilitative work. How can this work constitute making amends if it is not done by the actual perpetrators in question? Without actively engaged reparation from those most responsible for animal abuse and exploitation, where does this leave the professionals and volunteers who do the feeding, monitoring, and caretaking? Should we even expect them to see their caretaking *as* amends?

Palmer offers one possible solution to the problem of professionalized amends in interspecies repair, which is that for many animals, it is psychologically irrelevant who does the reparative work (2010, 104). Meanwhile, for other animals, the psychological considerations actually cut in the opposite direction, such that the human perpetrator of an animal's abuse is the exact wrong person to care for that animal, no matter how remorseful and eager for redemption they might be. There is nevertheless reason to expect and demand apology and amends of some sort from those directly responsible for animal abuse and exploitation and to criticize individual and corporate actors who treat the aftermath of their interspecies wrongdoing as nothing more than a budget line, even in cases where nonhuman victims cannot register apologies as such or where the actual reparative caretaking work is better performed by others. This is because human perpetrators of animal abuse and exploitation stand in strained moral relationships not only with their nonhuman victims but also with the rest of the moral community, the human members of which have a reparative duty to hold one another accountable for our actions. In this way we can see the work of animal advocacy as requiring both active bystandership and self-criticism, since each of us in some way or another has been both witness to and responsible for interspecies wrongdoing.

INTERSPECIES REPAIR AND SECOND-ORDER IMPLICATIONS

"If I had been in a position to design and create a world," writes Jeff McMahan (2010), "I would have tried to arrange for all conscious individuals to be able to survive without tormenting and killing other conscious individuals. I hope that most other people would have done the same." McMahan then proceeds to ask whether (and why not) we should bring an end to all predation, not just hypothetically but as active interventions in the actual world. His framing is illustrative of an ideal-theoretic ethic, seeking as it does to make sense of what our moral obligations *are* to nonhuman animals by reference to what these moral obligations *would be* if we had created them (and the rest of the world) from the ground up.

"But this is not the relation in which we stand to the other animals," as Christine Korsgaard (2018, 1186) reminds us. "We are not their creators, and we are not creating a world from scratch. We are the inhabitants of a world we already share with other animals, and the question we are asking is what we owe to *them*." What I appreciate about the varied accounts by Donaldson and Kymlicka, Gruen, Palmer, and Emmerman is that each of them takes this sentiment to heart and builds upon it in a meaningful way, whether in terms of interspecies political theory, entangled empathy, duties of assistance, or reparative practices. We are not creating a world from scratch, and the world we already share is one in which we have hurt, exploited, and otherwise wronged other animals—repeatedly, historically, and persisting today. To ask then what we owe them in *this* world cannot be limited to what we should do understood as what we should have done nor what we should stop doing understood as what we should never have done in the first place, as valuable as these counterfactual queries might truly be. Given the extant reality of interspecies wrongdoing, what we *now* owe other animals should include ameliorative duties of interspecies corrective justice.

What form our amelioration should take is a difficult question. Exclusively forward-looking abolitionist visions seem to neglect the need for justice in the unjust meantime (Jaggar 2019). Like Emmerman, I am more sympathetic to human-animal relational repair than a compensation model of interspecies restitution, yet it is not easy to get a clear sense of what reparative justice looks like for human-animal relationality. When are interspecies trust and forgiveness realistic goals? Does it make sense to talk about repairing animals' moral standing, such that we human perpetrators hold each other accountable for the wrongs we do them, or should we simply focus on skillful, experienced, affectively engaged *care* and let concerns about interspecies atonement and accountability drop away like so much unneeded moral ballast?

However we decide to answer these questions, we must also reckon with second-order moral implications that interspecies repair can have for human-human and interspecies relationships. Recall, for example, the Anishinaabe restoration program for depleted lake sturgeon populations in the Great Lakes, previously discussed in chapter 3. As Holtgren et al. explain, this restoration program has made significant progress in interspecies relational repair and also human relational repair between Anishinaabek and settler Americans in the region. "The fish has been able to both heal old wounds and create new, sustainable relationships among people, even in a watershed where these relationships have been strained by settler colonialism" (Holtgren et al. 2015).

There is a perennial need for relational repair not just between humans and other animals, but also among different human beings with different cultural or ethical commitments. We inevitably do moral damage to our various relationships that cut across these differences even as we do our (inevitably imperfect) best to balance integrity and respect, not just hypothetically but actually as we live, love, work, and eat together. Things can get complicated quite quickly. The perpetration of second-order harms given conflicting environmental ethics extends to something as politically fraught as the 2001 arson case at the University of Washington (in which monkey-wrenching activists set fire to academic buildings and destroyed unrelated scholarly research in the process), renewed Makah whale hunting, and recent US Fish & Wildlife Service practices of killing barred owls to protect endangered spotted owls in the Pacific Northwest.[22] It also extends to something as personal as a principled vegan refusing to eat her grandmother's beloved chicken soup and a grandmother keeping her family traditions alive while respecting her granddaughter's steadfast ethical commitments.[23] "We must not simply go along with the traditions of our people when those traditions are morally problematic," Emmerman (2019, 85) argues; and yet, she reminds us, "Even when those practices are immoral, our turning away from them, challenging them, or attempting to shift them can cause a moral cost in important relationships."

NOTES

1. See also Aaltola (2005, 20–22).

2. See also Regan (1982, 1995, 2004b), Regan and Francione (1992), and Francione (2000, 2005, 2009).

3. "While there is disagreement over the validity of positive moral rights, there is unanimity concerning the validity of negative moral rights," Regan (2003, 25) argues. "This unanimity among these thinkers makes our work easier . . . our inquiry will focus on negative moral rights (henceforth 'rights' for reasons of linguistic economy)."

4. On animal rescue generally see Anderson and Anderson (2006) and Crisp (2000); on farm animal sanctuaries see Brown (2012), Jones (2010), and Stevens (2010); on pet sanctuaries see Glen (2001) and Zheutlin (2015); on chimpanzee sanctuaries see Project R&R (2019); on plans for a marine sanctuary see The Whale Sanctuary Project (2019).

5. See also Callicott (1988), Burgess-Jackson (1998), and Haraway (2008).

6. See, for example, Singer (1974), Matheny (2006), Linzey (2009), and Ryder (2011).

7. Singer (1990, Ch. 2); see also Ryder (1975), Rollin (1992), and Linzey and Linzey (2018).

8. Singer (1990, Ch. 3); see also Mason and Singer (1990), Norcross (2004), and McMahan (2008).

9. "Despite their differing foundational premises, utilitarian and rights-based accounts of animal rights have, to date, both focused almost exclusively on universal negative rights" (Donaldson and Kymlicka 2011, 261–62).

10. See for example Adams (1990), Warren (1990), Curtin (1991), Gruen (1993), Adams (1994), Gruen (1996), Gaard (1997), and Wyckoff (2014); Taylor (2017) offers a particularly powerful intersectional analysis of animal rights and disability rights.

11. On relationships and responsibilities, see also Goodin (1985) and Scheffler (1997).

12. Afterword to Gruen (2015, 100).

13. Marino et al. (2009), Emmerman (2014a), Donaldson and Kymlicka (2015), and Doyle (2017).

14. See also Bostock (1993), Marino et al. (2009), and Gunasekera (2018).

15. In contrast to animal captivity in zoos and aquariums, Marino et al. characterize sanctuaries as "the remedy" (2009, 27).

16. See Palmer (2010, 2011, 2013, 2018, 2019).

17. See also Burgess-Jackson (1998, 163–70).

18. My thanks to Karen Emmerman for pressing this point about creating internal dependencies in animals that are not domesticated yet not quite wild either.

19. See also Palmer (2013, 30); through her critical reading of Donaldson and Kymlicka (2011), Higgins (2019) argues for reparations to other animals given the harms we have done them.

20. See also MacClellan (2013).

21. On climate change and our duties of assistance to animals, see also Palmer (2013, 2019), and Pepper (2019).

22. On the University of Washington arson case, see Liddick (2006) and Bernton and Clarridge (2006); on Makah whale hunting, see Gaard (2001), D'Costa (2005), and Kim (2015); on the barred and spotted owls, see Livezey (2010), Diller (2013), US Fish & Wildlife Service (2013), Moore (2016, 204–12), and Le (2019); on conflicting duties not to harm and duties to assist nonhuman animals, see Abbate (2016).

23. Emmerman (2019).

Chapter 5

Climate Change and Intergenerational Reparative Justice

Theories of intergenerational justice that speak to asynchronous obligations too rarely address what to do when we fail in these obligations. Ideally, we would never do future persons wrong, but when we do, what direction might we find in an approach to intergenerational injustice in terms of relational repair? In this chapter, I offer a reparative approach to intergenerational justice which brings together Walker on reparative justice and Baier on cross-generational relationality. Walker offers us tools for understanding and acting to restore relationships in the aftermath of injustice, while Baier considers the moral implications of membership in a cross-generational community of past, present, and future people, contrary to atomistic theories of intergenerational justice in philosophical debates on climate ethics today. Given injustices, how can perpetrators, victims, and cross-generational communities go forward; how can we repair relationships that span across generations? I argue that practices of intergenerational justice, including apology and acknowledgments of wrongdoing and making amends for asynchronous climatological damage, can strengthen the normative basis and give meaningful direction for cross-generational projects of climate justice.

(UN)KNOWINGLY ROLLING COAL

When Volkswagen admitted to using defeat devices to artificially reduce emissions as measured during vehicle testing, affecting eleven million cars sold 2008–2015 worldwide, loyal VW drivers were shocked and dismayed. Unbeknownst to them, their personal vehicles had been emitting (in normal driving modes, contrasted with the fraudulent test-only mode) at four times the European limit and *up to forty times* the US standard of the Clean Air

Act. As journalist and VW driver Andrew Stoy (2015) puts it, "That we are all unknowingly 'rolling coal,' spewing exponentially more emissions into the atmosphere than we realized, and that Volkswagen was fully aware of the deception, carries a potent sting for those of us who believed the extra cost of a VW TDI vehicle was worth the fuel economy and emissions benefits."[1]

If Volkswagen should hope to win back the trust of thousands of customers, it will have to pursue a long process of relational repair. But Stoy and customers like him are not the only ones wronged by VW's massive fraud, and the customer relationship is not the only one damaged and destabilized by the company's environmental damage. Researchers have estimated the numbers of premature deaths in the United States and Germany owing to excess VW emissions, as well as increases in nonfatal respiratory conditions, health expenses, and other social costs (Barrett et al. 2015; Chossiere et al. 2017). In this and other such cases of corporate malfeasance with climatological effects, the prospects for relational repair will turn on how the company, customers, and other affected parties handle the aftermath of environmental wrongdoing.

Volkswagen's emissions fraud was especially brazen, but this was neither the first nor the last time that human beings have been rolling coal, knowingly or unknowingly spewing greenhouse gases into the earth's atmosphere to devastatingly compounding climatological effect. For some time, we were like children riding along innocently in the backseat, alert to what we were doing (raising animals, cutting trees, burning fuel) yet unaware of the unseen atmospheric impacts. For some time further still, we were like loyal VW drivers, at some level aware that our actions were producing greenhouse gases but mistaken, misinformed, in denial (or some combination of these) about the amount produced and the extent of their impact. In this time, some individuals, groups, and organizations have been like Volkswagen on a vast scale, actively constructing ignorance on global climate change and its anthropogenic nature.[2] Most recently, our position has become akin to VW drivers reckoning with the emissions-fraud reveal. Whether innocent or otherwise before, we can no longer plead ignorance, nor can we continue driving (flying, burning, cutting, eating, etc.) as we did before.

Yet continue on we do. This horrible inertia is a major point of emphasis in climate activist Greta Thunberg's bracing speech before the United Nations in September 2019. In addressing her remarks to world leaders, Thunberg is self-consciously speaking across not only international but also generational divides, and she puts this partially overlapping relationality to good use, as David Wallace-Wells (2019a) describes it, "waging a rhetorical war against her elders through an unapologetic use of generational shame." Indeed, her aim is not to inspire hope but exhort action. "You say you hear us and that you understand the urgency. But no matter how sad and angry I am, I do not want to believe that," Thunberg told the UN. "Because if you really understood the situation and still kept on failing to act, then you would be evil. And that I refuse to believe."[3]

Evil. Shame. In popular and scholarly discussions of global climate change, there is a growing recognition of not just its practical but its ethical significance. As the Intergovernmental Panel on Climate Change *Fifth Assessment Report* (2013) points out, "Ethical judgments of value underlie almost every decision that is connected with climate change." Environmental ethicists who study climate change most often do so as a matter of international-intergenerational justice, and these questions of justice are most often framed in terms of ideal theory. This inattention to nonideal theorizing for climate justice is peculiar, since notwithstanding their genuine disagreements on how best to understand intergenerational rights and obligations, climate ethicists today agree that contemporary institutions and governments have failed (and continue to fail) in *some* significant sense to protect these rights and meet these obligations however understood. Injustice typifies the international-intergenerational climatological state of affairs as it stands more than justice, and a near-exclusive focus on ideal theory in the ethics of climate change is at worst negligent, at best unnecessarily limiting.

Each of the contributions to Heyward and Roser's book *Climate Justice in a Non-Ideal World* (2016) loosens the presumptive ideal-theoretic constraints on climate-ethics debates in some way or another. As the editors note, John Rawls's influential distinction between ideal and nonideal theory in political philosophy is built on two key idealizing assumptions: *full compliance,* such that all individuals and institutions in the relevant society act in accordance with the principles of justice, and *favorable circumstances,* such as moderate rather than extreme scarcity (Rawls 1973, 245). "For Rawls, it was necessary to engage in ideal theory before relaxing the assumptions and beginning non-ideal theorizing" (Heyward and Roser 2016, 6). A clear example of ideal theory in climate ethics is the assumption of full compliance with greenhouse-gas egalitarianism: that is, in theorizing about the extent and scope of our moral obligations concerning global climate change, it's typical to begin by assuming that everyone limits their emissions to their fair share. This isn't remotely true, of course, but the idea is that we start there, *then* consider what changes, ethically speaking, if the idealizing assumption is removed. For example, when and if others fail to meet their climate responsibilities, Simon Caney (2016) asks, how if at all does this change the ethical landscape? Does their noncompliance mean complying parties should carry even heavier ethical burdens? Does it mean we should lower our standards for what counts as compliance? Should we invest theoretical and practical resources to induce higher rates of compliance?

These questions rarely arise or are even particularly intelligible for ideal theorizing on climate justice. But what I find notable is something that Caney's project and the many other worthwhile contributions to *Climate Justice in a Non-Ideal World* do not directly address. If and when we fail to meet our climate responsibilities, what sorts of ameliorative responsibilities follow from this moral failure? The question here is not only what should be done to make up the difference between full and partial compliance but also how to reckon with the (longstanding, persisting, international, intergenerational) moral failing itself. We need practices of corrective justice given climate change; specifically, we need deliberately restorative practices toward intergenerational relational repair.

ON INTERGENERATIONAL RELATIONALITY

Among those philosophers and political theorists who have addressed what to do given climate injustice, many otherwise worthwhile and insightful analyses still tend to neglect relationality. In considering the allocation of costs of environmental protection, for example, Henry Shue (1999) defends commonsense principles of fairness including our contributions to the problem and our ability to pay, yet the need to repair the moral conditions of our intergenerational relationships damaged by anthropogenic climate change is not considered. "The argument that rich countries should pay for climate abatement, because they are the most responsible for the problem of climate change is an argument about what philosophers call corrective justice," write Posner and Weisbach (2010, 100). They then proceed to argue against corrective justice in favor of a general welfarist approach to climate justice, yet take it for granted that compensation for climatological harms is the only relevant candidate for corrective justice, the only sense in which perpetrators of such harms might have a special responsibility. Repairing the moral conditions of relationships hurt by climate change and committing to international agreements such as the Kyoto Protocol and the Paris Agreement as part of an intergenerational reparative process are not even addressed. In surveying the relevance of past wrongdoing for intergenerational justice, Lukas Meyer's (2015) otherwise careful and thorough discussion likewise approaches the matter entirely in terms of compensation.[4]

Perhaps this inattention to relational repair given anthropogenic climate change is due in part to disagreement among theories of intergenerational justice on whether past, present, and future persons should be thought to stand in meaningful relationship with one another. For example, Edith Brown Weiss (1992) builds her theory of planetary rights on the idea that "all generations are linked by their on-going relationship with the earth." From here Weiss argues that violations of intergenerational obligations by one generation may well create further obligations for other generations. When generations are conceived as entirely distinct and independent, by contrast, it makes little sense why one generation's environmental failures may increase the ethical burdens reasonably placed on a totally different generation. Consider Steve Gardiner's (2009, 2011) pure intergenerational problem (PIP), which he calls the central problem for intergenerational climate ethics: for PIP, generations are ideally modeled as completely temporally nonoverlapping, and the causal and temporal asymmetry between earlier and later generations preclude any possibility of intergenerational reciprocity (2011, 166).[5] While the pure problem is of course an idealization, Gardiner sees it as useful in explaining the ethical challenge of intergenerational buck-passing in more realistic conditions as well (2011, 170–74). Weiss and Gardiner theorize climate justice differently owing in part to their respective emphasis on and abstraction from intergenerational relationality as explanatorily significant.

For her part, Baier does not argue for reciprocity in the transfer and receipt of goods across generations, but that question of reciprocity is not the only intergenerational ethical issue for her. What matters for Baier is that past, present, and future persons are taken to stand in continuous community: "As members of a cross-generational community, a community of beings who look before and after, who interpret the past in light of the present, who see the future as growing out of the past, who see themselves as members of enduring families, nations, cultures, traditions" (1981, 177). Here Baier builds on a long tradition in political philosophy, from Edmund Burke's cross-generational community to John Rawls's social union across generations, and continuing on with Avner de Shalit (1995) on transgenerational communality and Janna Thompson (2009) on intergenerational justice.[6]

It is significant that this moral relationship includes past as well as present and future persons. Contrast Baier's cross-generational community with Brian Barry's account of intergenerational justice as strictly "justice between the present generation and future generations" (1999, 107).[7] Our duty of sustainability "to leave as much and as good of the public goods previous generations have bequeathed," Baier (1981, 176) argues, "rises as much from a right of past persons to have their good intentions respected as it does from any right of future persons." Notice that the rights of past persons are not limited to the past, then, but are also future-oriented in the sense that they impose what Lukas Meyer (2006, 413) calls *surviving duties*: "The rights imply duties that are (also) binding after the death of the bearer of the right if the appropriate bearer of the duty is identified." The assumption is not that past persons are now in any position to *know* present and future persons will respect their wishes, keep our promises, or continue intergenerational projects that mattered to them. Even if they cannot know, even if they cannot be made happier or sadder, disappointed or satisfied, past persons can nonetheless be indirectly affected by what present and future persons do (and what we do not do). Meyer explains the asynchronous moral relationality involved in acknowledging historical injustices as follows:

> In acknowledging past people as victims of egregious wrongs we cannot affect their well-being. Also, such acknowledgement cannot be expressed vis-à-vis the dead victims, but only vis-à-vis currently living people in light of the wrongs past people suffered. However, if it is true that we stand under surviving duties toward past victims of historical injustice owing to the wrongs they suffered, then our fulfilling the duty by publicly acknowledging the past injustices they suffered will change the relation between us and the dead victims of historical injustice. It will be true of the past victims of these injustices that they have the posthumous property that we fulfilled our surviving duty toward them. To be sure, a change of the relation between a currently living person and a dead person does not bring about or rely upon a real change of the latter person. Rather the relational change is based upon the real change of the person who carried out the act. (Meyer 2006, 415–16)

Consider a simple example: a woman dies before her only child has his own first child, and so posthumously she becomes a grandmother. She may not know this, but it is nevertheless true of her. Similarly, past persons may be victims of historical injustice and targets of intergenerational reparative justice, contingent upon present and future persons' reparative practices. To the extent that present and future persons fail to respect the rights of past persons as members of our cross-generational community, past persons may also be victims of fresh intergenerational injustices and so targets of intergenerational reparative justice in that respect as well.

Expanding her sense of generational interrelationality even further, Baier holds that because interdependency is transitive, our cross-generational community thus extends cross-culturally as well.[8] The full relational picture here is truly global, and strikingly different from most theorizing on intergenerational climate justice today. "To sum up, the chief facts are our indebtedness to the past and our dangerously great ability to affect the future," Baier says:

> We, like most of our forbearers, are the unconsulted beneficiaries of the sacrifice of past generations, sometimes seen by them as obligatory, often in fact nonobligatory. If we owe something in return, what is it, and what can we do for those who benefited us? The most obvious response is to continue the cooperative scheme they thought worth contributing to, adapting our contributions to our distinctive circumstances. (Baier 1981, 179–80)

Like Weiss, Baier recognizes that the space after intergenerational ethical failure requires ethical attention and finds that our failures to meet obligations to future generations thereby create nonideal obligations to them. Specifically, she says, "We incur obligations to compensate our victims in a future overcrowded world for the harm we have thereby done them" (1981, 177). Identifying this as a duty of compensation, Baier diverges from the reparative approach I take in this book, although our approaches still share an emphasis on relationality and interdependency. With that difference in mind, let's consider the benefits of reparative justice compared to compensatory justice for anthropogenic climate change.

CLIMATE CHANGE, COMPENSATION, AND REPARATIONS

Reparative environmental justice can make several useful contributions to how we understand the aftermath of anthropogenic climate change. First, as Emmerman warns us for inter-animal wrongdoing, framing intergenerational justice as compensation and treating intergenerational wrongs as something to be compensated invites a risky and self-serving sense of moral relief. Perhaps when all is accounted for and adequate compensation has been made, we can convince ourselves that we avoided doing anything wrong. It will be tempting for perpetrators to regard the comparable good brought about as balancing and cancelling out the preceding wrongdoing. Reading compensatory justice as successfully erasing injustice can be especially hard to resist for anthropogenic climate change, where the climatological effects of our carbon emissions may not be felt for decades. Earlier generations trying to account for how their wrongful actions affect later generations will generally do so in an interim between the actions and their felt effects. Rather than victims experiencing an injustice and then experiencing compensation or reparation for it, climate change compensation or reparations might precede or run concurrent to and thus seemingly obviate the felt effects of climate injustice. As such, perpetrators may be tempted to see this chain of events as catching what *would have been* a climate injustice just in time, so to speak. Reparative intergenerational justice by contrast does not mask the fact of climate injustice from either future generations' or present perpetrators' recognition, nor does it encourage either future generations or present perpetrators to forget our histories of climatological wrongdoings. The reparative aim is forgiveness, not forgetting; contrition, not compensation; renewed trust, not reimbursement. The work of intergenerational moral repair is dynamic and ongoing, just as intergenerational relationships are themselves dynamic and ongoing.

Consider how contemporary discussions identify mitigation, adaptation, and compensation as pillars of climate policy, three kinds of responsibility for addressing climate change.[9] The first includes strategies to reduce greenhouse-gas sources and increase greenhouse-gas sinks. "There is a consensus among climate scientists that mitigation alone is insufficient," says Simon Caney:

> Given the volume of greenhouse gases that have been emitted, we are already committed to some climate change. In virtue of this, there is an ethical imperative that we—humanity—take steps necessary to ensure that any climate changes that occur do not undermine what people are entitled to do as a matter of justice. In the terminology employed by climate scientists, "Adaptation" is required. (Caney 2012, 257)

Caney identifies inoculations against infectious diseases, designing new drainage systems, and building sea walls to protect against rising tides as examples of adaptation. Such actions do not reduce greenhouse-gas levels or change their atmospheric effects, but when successful they stop (some of) the harmful implications. Compensation comes into play when mitigating and adaptive measures are insufficient, and anthropogenic climate change does indeed prevent people from, as Caney (2012, 258) puts it, "doing that which they are entitled to do."

My main concern is not about this taxonomy of climate responsibility, but how we understand it on a restitutive model of justice. If corrective justice is about restoring material conditions to their state prior to injustice or providing an adequate compensation for such material changes, these pillars of climate policy are still instrumentally important but become unrelated to any admission or acknowledgment of previous and persisting wrongdoing. Rebecca Buxton (2019, 1999) argues that one thing that makes climate reparations different from (and for her purposes, preferable to) climate compensation is the issue of who can deliver them. Where compensation could be provided to a victim by a third party, reparations by their nature must be made by those responsible for the harm in question. I would add that, in cases in which those responsible are as yet unable or unwilling to apologize and make amends for their wrongdoing, third parties may play a much needed role in recognizing the wrongdoing, providing aid, and reaffirming victims' moral standing to call for accountability. But as important as such third-party measures can be, Buxton is quite right that they are not enough for reparative justice. They do not relieve those responsible for harm from their ameliorative duties, and they do not repair the moral relationship between perpetrator and victim that has been damaged or destroyed.

Restitution as compensation is additionally problematic for global climate change because of how entangled our partial contributions are in their collective ecological effects. What the debate between Walter Sinnott-Armstrong and his critics[10] on the ethics of individual emissions shows us is that neither individuals' nor groups' emissions function in isolation. Rather, other parties' contributions significantly affect how harmful or benign one's own contribution will prove to be. It is exceedingly difficult if not impossible to determine the comparable compensation owed by individual or group emitters from an earlier generation to members of future generations (not to mention what may be owed to past persons) considered in isolation. For reparative justice, by contrast, the moral response to climate injustice is not compensation but rather the accountability and amends needed to renew the moral conditions on which our cross-generational relationships may be rebuilt. To the extent that intergenerational amelioration is appropriate, it is not about achieving comparable compensation. Indeed, the very aspiration to determine an appropriately compensatory amount of intergenerational repayment would itself contradict the humility and accountability needed for intergenerational reparative justice. What we are doing instead in the work of reparative climate justice is making amends toward cross-generational trustworthiness and forgiveness.

A reparative approach to climate justice allows for and gives direction to relational repair for those relationships damaged indirectly by the parties' respective perpetrations and experiences of climate change. As Gardiner and Hartzell-Nichols (2012) recognize, the "temporal diffusion" of climate change presents distinct ethical challenges. As public understanding begins to recognize the anthropogenic nature of contemporary climate changes (frequency and severity of hurricanes, rising temperatures) it can be tempting to blame contemporary emitters for the attendant harms that humans, other animals, and the rest of the world are experiencing. Yet those most responsible for the climatological changes felt now might be long dead, and those most affected by current greenhouse-gas emissions may not yet be alive. Unlike the ideal-theoretic constraints of the PIP, this is not exclusively so: the asynchronicity of emissions and their effects is not *so* protracted as to preclude some temporal overlap between those who caused and those who experience particular climatological harms. This is part of the power of Thunberg's UN address: our world leaders are old enough and she herself is young enough that, temporal diffusion of climate change notwithstanding, in this case a future victim is able to confront her assailants.

While they may be the most causally responsible, reparative justice for anthropogenic climate change need not be limited to members of the eldest living generations. Others of us should also work to repair the damaged moral conditions of our relationships with those currently feeling the harmful effects of climate change, even if our current emissions do not cause the climate changes they are experiencing. (As with other instances of moral repair, this could include a relationship to oneself.) Consider by analogy a father who is not sexist toward his own daughter though he is sexist toward other women and girls; meanwhile his daughter experiences sexism from other men and boys. This father cannot simply claim that his sexist behavior toward others is irrelevant to the moral health of their father-daughter relationship, not when she is a victim of the very sort of injustice he himself perpetrates.[11] He doesn't necessarily need to compensate his daughter for this, but he must at least acknowledge his wrongdoing and depending on her particular priorities and perspective, do what he can to rebuild morally healthy trust between them. Participating in anthropogenic climate change similarly undermines morally healthy relationality with those who experience the harmful effects of climate change. Our respective experiences may not be directly causally connected. I may not have harmed you, and so I do not owe you compensation, but I am responsible for the very sort of thing you are experiencing. This calls for relational repair.

A reparative approach to intergenerational climate justice prioritizes interdependence and collaboration. Trust enables us to do together what we cannot alone, so when trust is damaged or eroded through betrayal, corruption, or other wrongdoing, our relationships cannot sustain the same collective action. Anthropogenic climate change is one case of collective action where the presence or absence of trust and trustworthiness enables or impedes coordinated international-intergenerational policies of emissions reduction, mitigation, and adaptation. To the extent that climate justice requires collaboration, and collaboration requires resilient relationships to sustain it, a reparative approach to climate justice offers theoretical and practical tools that are relevant to the moral health and resiliency of our intergenerational relationships.

VICTIM IDENTIFICATION, APOLOGIES, AND AMENDS

Understanding intergenerational justice in terms of relational repair means taking seriously victims and their relationships involved in climate change. Some might wonder whether the challenge of victim identification across generations is simply too great. Victim subjectivity is prioritized on Margaret Urban Walker's model of reparative justice, which gives normative force and direction to our reparative practices, but it also presents difficulties when this is extended to asynchronous relationships, as in the case of global climate change. As Sarah Fredericks (2019, 147) cautions, "The scale of climate change and the diffuse nature of its perpetrators and victims complicates both apology and forgiveness. Who exactly is to apologize for what to whom?"

For her part, Fredericks worries that individual-to-individual apologies for climate change are likely just to be swamped by the rest of humanity; she is more optimistic about collective-to-collective apology "because it can acknowledge complicated systems while preserving intentional agency, established relationships, and a feasible scale of action" (2019, 155).

One challenge of victim identification concerns the construction of injured parties *as* victims. Although the intention is not to disempower the individuals and groups who have been wronged, Haalboom and Natcher (2012) warn that for populations deemed vulnerable to climate change, *victim* and *vulnerable* are hazardous labels. Such appellations can obscure victims' specificity: their particular context, their distinctive forms of knowledge, even their complicity in ecological and climatological degradation. Their concern is that victim appellations disempower so-called victims and instead empower wrongdoers as the primary agents of reparative action.

Whether the language of victimhood serves to disempower is a challenge for intergenerational reparative justice not easily dismissed. Yet the need for reparative justice to center on victims' experiences and values serves as a safeguard against victims' marginalization, prioritizing their actual subjectivities in the planning, implementation, and sustenance of reparative practices. The efficacy of this safeguard turns on the accessibility of said victims' subjectivity, however, which raises a second issue for intergenerational justice. The asynchronous nature of intergenerational relationships is an epistemic impediment for the requirement that acts of amends be directed in accordance with victims' preferences and subjectivities. For one, present persons' knowledge of future persons' identities is a recognized challenge of intergenerational justice. To the extent that amends are made by members of earlier generations to members of later generations, the former may be unable to access the latter's subjectivities in order to direct their reparative work. Earlier generations may be inclined to substitute and appeal to their own subjectivity through misplaced if understandable empathy, giving the impression that perpetrators have deferred to future people even as their actual subjectivities remain inaccessible to them.

If we recall Baier's observation that the victims of present injustices and other wrongdoings include past persons *and* future persons as members of the cross-generational moral community, however, we realize that the work of repairing the moral conditions of that community is also directed by the subjectivity of past persons whose perspectives, preferences, and values might be better known to us. Furthermore, the inaccessibility of future victims' subjectivities itself might be incorporated into intergenerational reparative justice as a methodological principle, a caution for epistemic humility. Among other things, this methodological humility means pursuing open-ended, deliberately incomplete practices of making amends, which allow for and actively invite future generations' revision and adaptations of these ameliorative practices as they see fit.

INTERGENERATIONAL TRUST AND FORGIVENESS

Greta Thunberg (2019) concluded her remarks at the United Nations with a cross-generational warning: "You are failing us. But the young people are starting to understand your betrayal. The eyes of all future generations are upon you. And if you choose to fail us, I say: we will never forgive you."

Extending repair to intergenerational relationships presents several challenges for trust and forgiveness. If we follow Baier in recognizing our obligations to future persons as also in part obligations to past persons, this may provide a partial solution to the practical challenge of prioritizing victim subjectivity. In so doing, however, we add to the challenges of asynchronous relationality. For an account of reparative justice that prioritizes the fact of victim forgiveness, moral priority is given not to contrite wrongdoers' opinions about when they deserve forgiveness nor third-party opinions about when forgiveness should be given, but actual victims' decisions to give or deny forgiveness. Walker (2013b) argues against third-party forgiveness on the grounds that it fails to properly recognize the moral burdens that wrongdoing and injustice put on victims specifically, which even well-meaning bystanders do not share and thus cannot forgive.[12]

If past generations are recognized as among the victims of climate injustice, the fact of causal asymmetry seems to preclude their forgiveness and renewed trust in present or future persons as part of the reparative process. For an earlier generation to forgive a later one seems to require that their terms of forgiveness be articulated in advance, with no flexibility in responding to later generations' acts of apology and amends or lack thereof. Earlier generations could conceivably give detailed parameters of hypothetical forgiveness, so that if such-and-such acknowledgments were made and such-and-such amends were done, forgiveness would be warranted. But this awkward approach would require not only highly prescient anticipation of the particularities of warranted forgiveness but also highly prescient anticipation of the specific injustices that later generations might perpetrate, for which forgiveness could be cross-generationally extended.

The second part of this challenge concerns forgiveness or lack thereof from future persons. One option is to abandon the requirements of trust and forgiveness in cases of intergenerational injustice where future generations' trust and forgiveness cannot be ascertained. One drawback of this option is its marginalization of victim subjectivity, which otherwise serves as a significant source of normative force and direction for this model. Another option is to keep criteria of trust and forgiveness but understood in less literal terms, like how we might model an ecosystem's forgiveness of human degradation as the point at which migratory and local species repopulate a once degraded, now restored habitat. But future persons are persons, while wetlands, forests, and prairies lack subjectivity in a literal sense, so this second possibility might not properly prioritize victims' subjectivity. On reflection, I think the best approach here is one that keeps future victim forgiveness at the heart of reparative intergenerational justice, but gives up the expectation that repentant offenders should be able to know that they have been forgiven and trustworthiness has been established to victims' actual satisfaction.

The third challenge of intergenerational forgiveness builds on the last two: might hypothetical forgiveness be a better measure of intergenerational repair than actual forgiveness? An emphasis on actual victim forgiveness would seem to invite confusion. Climate injustices rarely if ever have a lone victim, after all. With multiple victims, having different priorities and subjectivities, some forgive more quickly or slowly than others. Is the most hesitantly forgiving victim to be prioritized? The possibility of differential forgiveness raises the issue of unreasonable reproach. In cases where victims are insensitive to genuine contrition from perpetrators, is it reasonable that victims have full authority to determine what relational repair looks like?

The issue of relationality puts these challenges of intergenerational trust and forgiveness on a different footing than, say, restitutive or retributive approaches to intergenerational justice. If the issue were remuneration then we could understand how actual victims may sometimes misjudge how much they are owed, in that hypothetical assessments might prove more reliable than their actual assessments. We could see how wrongdoers might meet their obligations whether actual victims appreciate that fact or not. But this leaves actually existing relationships obscured. Why should we assume a common threshold for perpetrator forgiveness applicable across different victims, given that each is rendering an assessment of particular relationships rather than, say, assessing the perpetrator's intrinsic moral worth? To prioritize victim subjectivity is not to assume victim infallibility. In actual fact, of course victims sometimes forgive more easily or hesitantly than would be healthy for our strained relationships. But the forgiveness and renewed trust vital for morally healthy relationships do not happen just because victims *should* forgive or trust again, any more than contrition or renewed trustworthiness happen just because perpetrators *should* be contrite or trustworthy. To displace the actual victims' forgiveness for hypothetical forgiveness would be to attempt relational repair without regard for the actually existing parties to the actual relationship. It would be to attempt to affirm victims' standing to call for accountability in their communities without regard for the importance of their voices actually being heard.

RECOGNIZING WRONGFUL INTERGENERATIONAL REPAIR

Acknowledging our actions as wrong and accepting responsibility for them are the minimal conditions that must be met to even start to set things right with victims of injustice; for actions to count as making amends, we must admit our wrongdoing and present the action as a redress for it. While the work of intergenerational climate justice can be done this way, however, when we consider climate mitigation and adaptation projects they are often relentlessly forward-looking, with little or no sense of acknowledgment and admission of wrongdoing. The space between climate mitigation and adaptation on the one hand and admitted intergenerational wrongdoing on the other is perhaps most glaring when our harmful practices carry on. How can we claim to acknowledge our ecologically destructive practices *as wrong* and yet continue to do them?

This challenge of unacknowledged wrongdoing and persistent climatological degradation can be seen as a problem for a reparative intergenerational justice, but I think it's better understood as a problem for business-as-usual climate mitigation, adaptation, and management policies or practices. Intergenerational repair can include arrangements in which third-party members of a community work to make amends, even as those in the past, present, or future who are directly responsible for anthropogenic climate change fail to admit their wrongdoing, change their ways, or demonstrate renewed trustworthiness. In this way, conditions for morally healthy relationships may be partially restored in an intergenerational global community without relieving responsible, unrepentant individuals or groups of their own reparative obligations.

Yet is relational restoration even an appropriate goal in light of serious and persisting climate injustices? Intergenerational relational repair might be criticized for two reasons: on the grounds that relational repair might conflict with other measures for intergenerational justice, and on the grounds that some relationships simply should not be repaired. Let's consider the first of these first. We might wonder whether the process of repairing a relationship will directly or indirectly serve to damage other relationships, or alternatively, whether the process of relational repair will hinder or compromise prospects for restitutive or retributive responses to injustice. It does seem quite possible that practices of intergenerational reparative justice can diverge from and conflict with practices best suited for intergenerational justice as understood on different terms.

That said, there is surely room for convergence. As emphasized above, while compensation is not itself the end goal, it could still play a significant role in a process of making amends toward relational repair. Reparative acts in making amends are not about repayment, but if perpetrators genuinely acknowledge their wrongdoing and recognize themselves as morally accountable to their victims, accountability requires addressing how they have wronged their victims. The key difference with compensatory versions of corrective justice is that compensation is not itself the goal but part of a process by which perpetrators can rebuild trustworthiness, victims may extend forgiveness, and communities reaffirm the moral standing of their members.

How might the repair of one relationship serve to damage or threaten another relationship? Given the intertwining nature of intergenerational relationships among various individuals and subgroups in the past, present, and future, it is not difficult to imagine. Seemingly contrite wrongdoers owning their responsibility for one injustice while ignoring or disregarding their responsibility for another can serve to further erode the moral standing of the unacknowledged victims of the latter injustice, as their lack of success in calling for moral accountability is contrasted with the former. Alternatively, relationships among co-perpetrators of a collective injustice might themselves become strained when some but not all of them acknowledge their wrongdoing and seek to make amends toward renewed intergenerational trustworthiness and forgiveness. Relationships among victims of a collective injustice also might be strained by processes of relational repair if and when they disagree about why or when to extend forgiveness and renew trust. To recall Fredericks' observation noted above, the diffuse nature of perpetrators and victims of anthropogenic climate change complicates reparative justice. That said, refraining from the moral repair of a relationship can also damage or threaten other relationships, whether calling into question partially penitent perpetrators' trustworthiness or creating divisions among those whose moral standing in the community has been reaffirmed via repair and those whose moral standing has not. The wider lesson is not that intergenerational relational repair should be avoided, but that the processes of intergenerational relational repair must themselves be subject to moral scrutiny.

Subjecting processes of intergenerational repair to moral scrutiny means, among other things, considering whether some broken relationships should not be repaired at all, in which case their abolition rather than reform is what's called for. Abolition, in this sense, might be defended in at least three ways. First are those cases in which the perpetrators might rightly acknowledge their environmental wrongdoing and try to make amends to demonstrate renewed trustworthiness, yet future victims are still warranted in withholding trust or forgiveness indefinitely. In other cases, reparation is so unsuitable given intergenerational climate injustice that even the attempt to make amends would be inappropriate. In these cases accountability might mean freely and sincerely admitting fault and leaving be, and knowing that efforts at intergenerational reconciliation would be unwanted and unwelcome.

The third context is akin to what Kathryn Norlock (2010, 36) calls "rational pessimism": victims have no reason to expect perpetrators to admit their wrongdoings and every reason to expect them to remain unrepentant.[13] In the absence of any such reasonable expectation by victims of injustice that perpetrators will take responsibility, apologize, and make amends, repair may be beyond warranted aspiration.[14]

It does not contradict a reparative approach to intergenerational justice to acknowledge that the moral conditions of relationship sometimes cannot or should not be repaired. Abolition is a more viable option for some relationships than for others, however, when the cross-generational community itself is at stake. Recall Baier's (1981, 178) warning: "This power to end the human community's existence could justifiably be exercised only in conditions so extreme that one could sincerely believe that past generations would concur in the judgment that it all should end." That said, the cross-generational community can and should persist and be repaired even as some specific dysfunctional relationships cannot and should not. Perhaps Volkswagen should not repair its damaged relationships in the aftermath of its global emissions fraud, not because there is nothing to apologize for and nothing to be done, but because Volkswagen as a corporate entity should no longer exist.[15]

NOTES

1. See also Rogers and Sylvers (2015) and Mouawad and Jensen (2015).

2. On climate skepticism, see Singer (2000), Lomborg (2007), O'Keefe and Kueter (2010), and Michaels (2011); for critical evaluation of such practices, see Michaels (2008), Oreskes and Conway (2010a, 2010b), Ceccarelli (2011), Almassi (2012a), Gelfert (2013), and de Melo-Martin and Intemann (2018).

3. Thunberg (2019). As she said at the World Economic Forum at Davos in January 2019, "I don't want your hope. And I don't want you to be hopeful; I want you to panic. I want you to feel the fear I feel every day. And then I want you to act, I want you to act as you would in a crisis. I want you to act as if the house is on fire, because it is" (in Wallace-Wells 2019a).

4. On climate change and compensation, see also de Shalit (2011), Nine (2012), and Baatz (2013).

5. By contrast, see McCormick (2009) and Heath (2013).

6. Baier (1981, 177); see also Randall (2019).

7. Hiskes (2009) also takes a future-oriented approach to intergenerational justice.

8. In this way, Baier avoids Barry's (1999) critique of de Shalit's (1995) communitarian approach.

9. On the three pillars of climate policy, see Caney (2012), IPCC (2013), and Mintz-Woo (2019).

10. Sinnott-Armstrong (2005), Hiller (2011), Nolt (2011), Almassi (2012b), and Broome (2019).

11. Or consider two friends, a Black American who has experienced racist police violence and a white cop who has committed racist police violence although not directly against his friend. The cop cannot claim his violent acts are irrelevant to their friendship. This doesn't mean the cop owes his friend restitution or compensation, nor does his culpability erase or replace that of the officer who assaulted his friend, but their friendship will need relational repair nonetheless. (My thanks to Chris Partridge for helping to develop this example.)

12. On third-party forgiveness, see also Maclachlan (2008) and Norlock (2009).

13. See also Norlock (2017). Norlock emphasizes that pessimism and hopelessness need not imply despair and inaction: "When hope is all but gone, I require better reasons for action. Nonideal pessimism reminds me of the moral demands of degraded environments and vulnerable people, of the worth of the perpetual struggle" (Norlock 2019, 16).

14. On forgiveness and injustice, see also Card (2004), Schott (2004), and Holmgren (2012).

15. My thanks to Peter Steeves for pressing this point.

Chapter 6

Traditional Ecological Knowledge and Reparative Epistemic Justice

In this chapter, I expand the scope of environmental wrongdoing under consideration to include epistemic injustices—that is, wrongs perpetrated against individuals and groups in our capacities as knowers. Testimonial injustice, discursive injustice, epistemic othering, and hermeneutical ignorance are among the many varieties of epistemic injustice relevant to the fraught relationship between settler colonialism and traditional ecological knowledge (TEK). My analysis builds on multiple accounts of various epistemic injustices, Robin Kimmerer and Deborah McGregor on TEK, and Kyle Whyte on settler colonialism and indigenous governance. In this way, I seek to extend the model of reparative justice from the previous chapters into the epistemic domain, with specific attention to the prospects and challenges for repairing epistemic trust in the aftermath of environmental wrongdoing. In such cases, are renewed epistemic trust and relational repair possible or desirable? When does a reparative process demand too much or too little of the victims or perpetrators of epistemic injustice, and when does this process itself risk perpetuating second-order injustices concerning TEK?

CHURCH ROCK REVISITED

The Church Rock nuclear disaster and its aftermath were built on ignorance and inaction, where not knowing and not acting both mattered. As Church Rock residents and United Nuclear employees would tell journalists, historians, and government agencies, there were significant social-epistemic failures to be found prior to, immediately after, and in the long-term response to the July 1979 dam break. Consider, for example, the ways that UNC transmitted knowledge that the Rio Puerco had been contaminated. "In accordance with a state contingency plan, Navajo employees of United Nuclear Company were dispatched to personally notify Navajo-speaking residents downstream," write Brugge et al. (2007, 1598). "Within weeks, signs were posted in New Mexico and Arizona that warned against the use of water for human or livestock consumption." Navajo shepherd Tom Charlie confirms that UNC posted "contaminated wash" signs along the river. "But our cows, sheep, and horses can't read that. Most of us can't read, write, or speak English. The signs do no good," Charlie said. "It is wet now, but on days when it dries up, the wind comes along. The dust settles on the grass. The sheep eat it. We eat the sheep. We wonder what that does to our lives" (Wasserman et al. 1982, 151).

The radioactive contamination of the Rio Puerco raises important questions of knowledge and ignorance regarding water and livestock safety. With the main source of water polluted by millions of gallons of radioactive slurry, how much clean water was needed to replace it, and who was best positioned to know? As noted earlier, the Church Rock chapter of the Navajo Nation estimated a need of 30,000 gallons of clean water daily, of which UNC provided just a fraction. This discrepancy presents a dilemma where neither possibility is especially good. On one horn of the dilemma, UNC officials failed to grant sufficient credibility to the Navajo Nation to take their testimony seriously; on the other horn, UNC took the Navajo testimony seriously yet significantly underdelivered on the water needed. The first horn is an epistemic injustice, failing to extend a reliable and well-positioned knower the testimonial credibility that is warranted. The second is gross negligence, failing to give adequate assistance after a disaster of the company's own making.

Let's return to the question of livestock: were cows, sheep, and other animals grazing in the contaminated area safe to eat? Pasternak explains how New Mexico's official investigation and recommendations on the safety of Church Rock cattle exhibited a troubling ignorance of relevant Navajo practices:

> Reaux reported that the study "proved that health effects on humans eating animals grazing near/on the Rio Puerco/Church Rock area was not significant." But the researchers found markedly higher levels of radioactive nuclides in cattle there compared to livestock grazing in nonmining areas. Their advice was that eating the meat would not pose a problem *as long as the people there didn't depend on the butchered animals for their everyday food over a long period of time.* That, of course, is exactly what the Navajos did depend on. When they butchered an animal, they ate each one down to the bones, which they sucked around the fire. (Pasternak 2010, 151, emphasis in original)[1]

As with the water question, there are two possible explanations here. Both horns of the dilemma involve some degree of troubling disregard. The conclusion that the total health effects of eating livestock were insignificant was predicated on an ignorance of Navajo practices (and specifically practices directly relevant to the study itself), or the dismissal of such practices as unimportant or irrelevant. This is bad science, just as if livestock had been sampled only upstream of the point of contamination or researchers had presumed the local human residents were all strict vegetarians. To add insult to injury, downstream residents' requests for food stamps to supplement their lost livestock were denied (Wasserman et al. 1982, 151). As with the drinking-water contamination, authorities dismissed or disregarded local residents' testimonies about how the UNC disaster had affected their daily lives.

I said earlier that when Superfund site-designations were assigned in 1980, the UNC mill was designated while the adjacent, recently contaminated river was not. But if the signs posted just months before read "contaminated wash keep out," why not? Here again the failure was in part epistemic. As Pasternak (2010, 150) explains, Superfund laws required the EPA to establish causal responsibility for the specific environmental hazard in question. But no prior testing of the river had been done, so there was no baseline for comparison. The Rio Puerco was highly radioactive, to be sure, but how much of that was due to the July 1979 accident, how much due to spillage, seepage, or runoff from other mines over the years, and how much due to natural radiation in a uranium-rich environment? The EPA could not say and could not affix blame. Worth noting here are the epistemic standards prioritized and those that were not. Ideally the EPA would have had prior measurements of radioactivity for the Rio Puerco for many reasons, only one of which is to be a baseline for disaster comparison. Given the river's importance for human and nonhuman life in the area, and decades of uranium mining throughout the Southwest, the absence of measurements was itself a glaring instance of systemic ignorance. This not-knowing could have been passively allowed as a result of regulatory agency underfunding or understaffing; it also could have been actively maintained in the interest of certain parties.[2] A well-established baseline for comparison could either assign or absolve UNC from causal and legal responsibility, but with no established baseline, and ignorance maintained, where does this leave the work of ecological repair after the dam breach?

Consider the governmental response to ignorance about the river's greater radioactivity. In the absence of prior measurements, EPA was not authorized to alleviate ignorance by other means, for example, by working with the Church Rock chapter of the Navajo Nation or calling on other residents' historical knowledge of the river's health. The defense here cannot be that scientific knowledge of a river's health is more reliable than, say, oral histories and TEK, because in this case, the EPA had neither. Nor is the burden of overcoming ignorance put on the polluting party. To appreciate the burden of proof here, consider a contrasting evidential standard where a uranium-mining company would be expected to demonstrate that dangerously high levels of radioactivity are *not* due to its dam breach and radioactive spill. On *that* standard, maintaining such easily dispelled ignorance would run contrary the company's own interests and UNC would have good reason to keep reliable measurements of radioactivity in the Rio Puerco and elsewhere in the surrounding region.

We can see how the tail wags the dog. As Pasternak explains, a Superfund designation was predicated on affixing responsibility as determined according to an industry-friendly burden of proof, which holds not only for industry liability but also for allotment of governmental Superfund funds for cleanup if and when the responsible parties no longer exist. Establishing a baseline for comparison might be desirable, but not always essential to the question of environmental safety. Knowing *how much more* radioactive the Rio Puerco was after the UNC spill might be otherwise valuable information, but it is not strictly speaking necessary in order to know the river was too radioactive for human, animal, and ecological health.

Finally, consider the epistemic failures involved in the dam breach itself. As Wasserman et al. (1982, 148) concluded, "What made the Church Rock disaster especially tragic was that it could have been avoided." At a Congressional hearing on October 22, 1979, Representative Mo Udall argued that multiple regulatory agencies could have and should have foreseen the accident (US House of Representatives 1979, 3). A consultant had advised UNC before construction that *this* design on *this* site would be prone to settle and crack. In addition to that standing concern, at the time of the breach the dam was carrying a higher load than its design allowed for (Wasserman et al. 1982, 148–49). Navajo citizen and UNC employee Larry King was one of multiple workers who had observed and reported cracks in the dam before the accident, reports that UNC failed to relay to state or federal bodies (Gilbert 2019). "There were significant warnings appearing before the dam broke," concluded William Dircks, director of the Nuclear Regulatory Commission's (NRC) Office of Nuclear Material Safety and Safeguards. "I think that is the troubling part of it" (Wasserman et al. 1982, 149).

The issues of ignorance and burden-shifting for the Church Rock spill and cleanup contrast starkly with the subsequent decision to reopen the mill. Speaking on behalf of the Navajo Nation at the aforementioned Congressional hearing, Frank Paul argued that the mill should stay closed "until such time as a safe and sane method of dealing with uranium tailings is devised, tested, and implemented" (US House of Representatives 1979, 5).[3] UNC chief operations officer David Hann, by contrast, urged the need to resume operations: "We are concerned that continued denial of permission to restart our mill will force us to reduce our workforce substantially, resulting in severe hardship to the local community" (US House of Representatives 1979, 25). The NRC let the mill to reopen on November 2, "a process that led to widespread groundwater contamination and placed the United Nuclear Corporation Church Rock Mill on the Environmental Protection Agency's National Priorities List in 1983" (Brugge et al. 2007, 1598; see also US EPA 2020).

The UNC dam breach might remind some readers of WK Clifford's shipowner story in his essay on the ethics of belief. Clifford tells of a shipowner who actively avoids growing evidence of his ship's dangerous condition. He deliberately avoids looking too closely, finds various ways to dismiss the testimony of several concerned crew members, and leans heavily on the ship's past success as reason to believe in its future safety:

> In such ways he acquired a sincere and comfortable conviction that his vessel was thoroughly safe and seaworthy; he watched her departure with a light heart, and benevolent wishes for the success of the exiles in their strange new home to be; and he got his insurance-money when she went down in mid-ocean and told no tales. (Clifford 1886)

The larger lesson is that convincing oneself of something contrary to the available evidence is no defense of the irresponsible actions derived from that irresponsible belief. Instead, Clifford says, we should hold each other and ourselves responsible for our beliefs and our actions.[4] From gross negligence to stereotype biases, the evidential bases for our individual and institutional doxastic commitments call for normative assessment on epistemic and ethical grounds.

Some of the failures preceding, accompanying, and following the 1979 Church Rock nuclear disaster could perhaps be counted as epistemically unwarranted or inadequately justified but not unjust or unfair. For others, however, and indeed for many cases of ecological knowledge and ignorance, my suggestion is that these failures are best understood as committing one or more varieties of epistemic injustice: that is, as Miranda Fricker (2007, 1) explains, one or more ways in which epistemic agents are wronged in their capacities as knowers. And here I am particularly interested in injustices concerning TEK, ranging from willful neglect and general disregard to epistemic objectification and exploitation.

Before turning squarely to epistemic injustice and its relevance to the misuses and abuses of TEK, I should say a bit about what I mean by TEK. Beyond the initial gloss that TEK refers to knowledge of particular native and indigenous peoples concerning the natural world, there is no clear consensus among those who discuss TEK as to its conceptual content. Kyle Whyte (2013, 3) posits that different accounts of TEK can be usefully distinguished by two assumptions: knowledge mobilization, and the relation between TEK and science. The first of these concerns what different kinds of knowledge can be used for various purposes in various contexts. Some see TEK as a body of indigenous knowledge created, stored, and ready to be applied;[5] for others, TEK is not an archival store of accumulated knowledge so much as an ecologically situated way of knowing, a participatory activity or set of activities in which the knowledge and knowers are interrelated.[6] On the former conception, Whyte (2013, 5) says, TEK can be readily extracted from its society and plugged into policy-relevant science, while on the latter conception such knowledge cannot be rightly integrated into a new context of application unless the people who participate fully in it are also brought to the table as epistemic equals.

The relationship between TEK and science is similarly disputed. Some see the distinction itself as misleading and counterproductive, since good knowledge is good knowledge. "The attempt to create distinctions in terms of indigenous and western is potentially ridiculous," writes Arun Agrawal (1995, 433). For her part Robin Kimmerer sees opportunities for exchange and cross-pollination between traditional and scientific ecological knowledge (SEK) and a role for both in mainstream biology education. "Both knowledge systems yield detailed empirical information of natural phenomena and relationships among ecosystem components," Kimmerer (2002, 433) observes. "Both SEK and TEK have predictive power, and in both intellectual traditions, observations are interpreted within a particular cultural context." Kimmerer does not reject the SEK/TEK distinction completely: she sees them as complementary ecological perspectives that come from complementary but differing worldviews. For still others, TEK and SEK are totally distinct, incommensurate knowledge systems, perhaps because TEK is tied to spirituality and SEK is resolutely materialist, perhaps because SEK demands objectivity that TEK transcends. As El-Hani and Souza de Ferreira Bandeira (2008, 758) put it, "We are just saying that they are different, and should be kept different, for the sake of clarity about the nature of knowledge and the nature of science."

Whyte is not interested in identifying one best definition of TEK. He argues instead for understanding TEK as a *collaborative concept*, to "advance environmental governance even as we discuss and disagree on the assumptions underlying various definitions" (2013, 8). Whyte regards environmental governance today as at a significant inflection point: institutions such as the US Forest Service and Environment Canada, which were created at a time when the idea of indigenous environmental governance was completely marginalized, now show greater respect for indigenous knowledge and practices. "These changes create opportunities for indigenous peoples to work collaboratively with non-indigenous peoples, instead of against them or in secrecy from them" (ibid.). This makes TEK important *because of* disagreement over what the concept entails, not despite it, as "those who bring new definitions of

TEK into dialogue are inviting others to consider new possibilities for thinking about the function of knowledge systems in environmental governance" (2013, 9).[7] That TEK is a contested concept is part of its value for collaboration, if only because its meaning and proper usage have not been settled. Our appeals to TEK in cross-cultural collaborative contexts invite differently positioned parties to reflect on the different ways of conceptualizing it, rather than taking one definition for granted.

I hope to shed light on the relevance of epistemic justice to TEK in a similar fashion, where the contemporary contestedness of TEK allows us to reflect on how we wrong and are wronged by others in our capacities as knowers. In this way we might also appreciate the relevance of epistemic injustice and reparative epistemic justice to making sense of and valuing TEK in both colonialist and collaborative contexts of environmental governance.

EPISTEMIC INJUSTICES IN ECOLOGICAL CONTEXTS

Let us get a sense of the many varieties of epistemic injustice identified in the recent literature, with special attention to their relevance to TEK. In her work on *testimonial injustice*, Miranda Fricker identifies both credibility deficits and credibility excesses as epistemically errant. Both constitute a mismatch between a person's rational authority in some field and the credibility extended to them. As Fricker sees it, only credibility deficits are unjust (2007, 19); by contrast, Medina (2011), Davis (2016), and Lackey (2016) all argue that credibility excesses can produce testimonial injustices too.[8] Other injustices include *epistemic objectification* (Fricker 2007, 6; Tuvel 2015), in which a listener treats a speaker as a source of information but not as a genuine knower or epistemic subject in their own right; *trust injustice* (Marsh 2011), in which a speaker is not trusted owing to listener prejudices and biases; *participatory injustice* (Hookway 2011; Pantazatos 2017), in which a speaker's utterances are errantly rejected as discursively irrelevant; *argumentative injustice* (Bondy 2010), in which a listener puts too much or too little credibility in a speaker's argument due to a credibility excess or deficit; and what Kristie Dotson (2011) calls *testimonial quieting* and *smothering*, in which a speaker is prevented from testifying at all owing to a lack of listener uptake, either by quieting the testimony she is attempting to give or by smothering the testimony she otherwise would have given.

There are also injustices relating to speech and speech acts, including *interpretive injustices* (Peet 2015), in which speakers' utterances are misinterpreted in meaning due to listeners' bias, and *discursive injustices* (Kukla 2014), in which biases, stereotypes, or identity prejudices can undermine a speaker's ability to perform a particular speech act and instead compel performance of an entirely different speech act. For example, Cassie Herbert (2017) explains how survivors of sexual assault and harassment are disbelieved when their *reports* (which call for default belief) are misread as *accusations* (which call for default disbelief), which thereby undermines what survivors are trying to say and affects the pragmatic output of their testimonies.

Epistemic injustices can affect one's ability to speak, to be believed, and to be heard, but also one's ability to understand and be understood. The last of these is what Fricker (2007, 147) calls *hermeneutical injustice*, which in her formulation concerns an inability to understand one's social experiences owing to hermeneutical marginalization *and* an inability to get others to understand this experience.[9] Rebecca Mason (2011), Kristie Dotson (2012), and Gaile Pohlhaus Jr. (2012), by contrast, argue that these things can and do come apart. A victim of epistemic injustice and those relevantly like her may have the conceptual resources they need to understand their own experiences, even while others fail to understand because they do not avail themselves of those resources. Such failures may sometimes be merely accidental, but not for what Pohlhaus (2012, 715) calls *willful hermeneutical ignorance*; and when privileged persons' failures to understand lead to more systemic epistemic harms for marginalized knowers, this is what Dotson (2012, 30) calls *contributory injustice*. Particularly infuriating is the combination of willful hermeneutical ignorance and what Nora Berenstain (2016) calls *epistemic exploitation*, when oppressed persons are compelled to do the typically undervalued, uncompensated, yet indispensable epistemic labor that is expected and required to educate privileged persons about their oppression.

Across these many contributions to the burgeoning literature, various phenomena of epistemic injustice experienced and perpetrated have been identified and described in detail, and to a lesser degree, some attention has been paid to what should be done in light of such epistemic injustices. Let's consider how these analyses of epistemic injustices and their amelioration might usefully apply to a wide range of epistemic harm perpetrated against indigenous knowers and TEK.

Credibility Deficits and Excesses

It's not wrong to regard testimony with some degree of suspicion and critical scrutiny, and even when our reasons for suspicion are not especially strong, we have not necessarily wronged a speaker by extending them less credibility than rationality dictates we should have. When this mismatch between rational authority and credibility owes to some negative identity-prejudicial stereotype, however, this is what Fricker (1998) identifies as a core case of testimonial injustice. Clifford's shipowner is dangerously *negligent* in dismissing crew members' credible testimony on the ship's state; he commits *testimonial injustice* against these crew members if and when he gives their testimony less weight because of negative prejudicial stereotypes he applies to them, such as religious, class, and ethnic identity biases.

Testimonial injustice is among the most common epistemic wrongdoings committed against indigenous knowers. Examples abound in the TEK literature of European explorers, evangelists, settlers, and scientists dismissing tribal and aboriginal knowledge of ecological phenomena. For example, Nakashima (1993) compared the knowledge of Hudson Bay eider collected by wildlife biologists and Inuit hunters. The latter long had been dismissed as unreliable Eskimo reports, yet side-by-side comparison showed their knowledge of eider range, mortality, demographics, and behavior far exceeded that of the wildlife biologists. Credibility deficits are not limited to TEK testimony but extend to arguments and explanations as well. In *Braiding Sweetgrass,* Kimmerer recounts a difficult, ultimately rewarding process of directing a graduate thesis on the ecological benefits of harvesting grass for basketmaking. This project experienced academic pushback, not because it was costly or complicated, but because of its roots in TEK. Skeptical faculty members perpetrated argumentative injustice: the graduate student was not testifying to, but arguing for the scientific significance of her proposed project. At this initial stage she was not asserting that gathering half of a sweetgrass harvest has positive ecological effects, but arguing that this was worth studying. One faculty member rejected TEK as a theoretical framework; the dean called the project a waste of time because "anyone knows" harvesting a plant damages its population. But in the end, the study's results undercut the dean's presupposition and supported traditional practices. "Getting scientists to consider the validity of indigenous knowledge is like swimming upstream in cold, cold water," Kimmerer (2013, 160) writes.

It is relatively easy to identify cases of testimonial and argumentative injustice when listeners are upfront about their prejudices, and relatively harder to do when listeners cannot admit or even recognize these prejudicial TEK stereotypes in their views. After the nuclear disaster, for example, did UNC provide Church Rock residents too little water because they undervalued the Navajo Nation's recommendations or did UNC put enough credibility in their recommendations yet underdelivered on clean water anyway? When NRC let United Nuclear reopen the mill four months after the disaster, against the Navajo Nation's argument at the Congressional hearing, was it because NRC discounted the merits of this argument based on indigenous knowledge, or did the agency give the argument its proper due, epistemically speaking, then let the mill reopen anyway? In each case recipients of post-disaster testimony and argumentation might be guilty of unfair credibility assessments, negligent decision-making, or both.

When is extending speakers *too much* credibility not just unwise but unjust? Emmalon Davis argues that prejudicial credibility excesses due to tokenizing and typecasting present this issue. "Harm arises," she says, "when a marginalized speaker's acceptance in an epistemic community or inclusion in a testimonial exchange is conditional upon the speaker adopting a—the—voice of distinction" (Davis 2016, 490). So understood, the wrongdoing here is about *epistemic othering*, "through which the capacities of a speaker are prejudicially assessed in such a way that bypasses or circumscribes the speaker's subjectivity" (ibid.) Davis diverges from Fricker and others who emphasize the need for individual epistemic virtues like critical awareness and virtuous listening. Epistemic amelioration in which hearers attend to improving their own characters is inadequate, Davis (2016, 494) explains, if they do not also "improve their epistemic environments—in which marginalized knowers are all too often either conspicuously present or (in)conspicuously absent."

Unjust credibility excesses arise when Western explorers, settlers, or scientists recognize that TEK can be useful for their purposes yet fail to recognize indigenous knowers as persons with specific, unique subjectivities. Consider, for example, an individual tribal member granted credibility but only as a token representative of the tribe's collective knowledge, irrespective of their own epistemic strengths and weaknesses. Tokenized credibility excess might occur at a group level too, as when scientists start to appreciate TEK and yet still fail to appreciate the epistemic differences between different indigenous groups. An indigenous group might be tokenized by recipients who assume that their TEK is representative of TEK across all communities and ecologies.[10] In both individual and group cases of tokenizing and typecasting, nonindigenous scientists and other recipients of indigenous testimonies should be assessing a speaker's own credibility and capacities, rather than prejudicially typecasting them as a generic indigenous knower. But as Davis will remind us, working on one's own individual awareness is only part of the solution. We must also scrutinize and improve our social-epistemic environments. Who is excluded from constructing our communities' evidential standards and collective epistemic resources? Who is allowed to participate in their construction and ongoing revision? Who is granted the social-epistemic authority to determine who may be excluded from or allowed into this process?

Participation and Exclusion

Hookway describes participatory injustice such that a listener errantly dismisses a speaker's utterance as irrelevant to the collective epistemic task at hand, which could be a public hearing, a conversation, or as in his main example, a classroom discussion. "The student is wronged, not because the teacher refuses to give credence to his testimony," Hookway (2010, 151) explains, but rather "because the teacher refuses to recognize him as a participant in debate or discussion." Credence is not the issue: indeed, by dismissing the student's voice as discursively irrelevant, the degree to which his testimony is or is not credible is simply put aside. Participatory injustice has its own distinct harmful implications, epistemically and practically speaking. Unlike testimonial injustice, participatory injustice cannot be ameliorated by revising our credibility deficits up or excesses down, given that the speaker's potential contributions were ignored altogether.

Hookway cautions that many of our epistemic practices already presuppose the possibility (or impossibility) of our participation as hearers, critics, and fellow inquirers. "We must be critically aware of the way in which our judgments of the intended relevance of contributions can be influenced by stereotypes and prejudices" (2010, 151). Though judgments of relevance cannot be eliminated, critical reflection can help us to focus them and enable us as listeners to better see the relevance of contributions we would have otherwise dismissed. This critical reflection extends to past judgments too, particularly judgments that perpetrated participatory injustice. Sometimes our belated recognition of the discursive relevance of a speaker's contributions and the speaker as a full participant proves better late than never. But the distinct epistemic harm of participatory injustice also means that testimony once excluded is less likely to be included in the conversation later, since the speaker has already been sent the message that their voice is distracting, derailing, or irrelevant.

As noted in the prior section, Robin Kimmerer argues for incorporating TEK into biology education for multiple reasons, including SEK/TEK cross-pollinations but also to make mainstream science less hostile to and alienating for native students (2002, 435). In graduate school, Kimmerer found herself teetering precariously with one foot in a scientific world and one in an indigenous world. "To walk the scientific path I had stepped off the path of indigenous knowledge," she writes. "But the world has a way of guiding your steps" (2013, 44). She learned something simple but remarkable, that an American elm, the largest of its kind, had been named for her Potawatomi elder Louis Vieux. Learning about this elm was the beginning of her journey back to her people. Everything about graduate education told Kimmerer that she had to choose between indigenous and scientific knowledge. "But then," she says, "I learned to fly":

> Or at least try. It was the bees that showed me how to move between different flowers—to drink the nectar and gather pollen from both. It is this dance of cross-pollination that can produce a new species of knowledge, a new way of being in the world. After all there aren't two worlds, there is just this one good green earth. (Kimmerer 2013, 47)

Kimmerer envisions a world in which native students participate as full members in collective scientific inquiry without having to give up their connection to and participation in indigenous knowledge and cultures. Otherwise native students and scientists will be forced to smother their valuable TEK-based testimony or they will be further excluded from participating in collective SEK practices.

Discursive injustice and interpretive injustice are other ways in which indigenous knowers are excluded from adding to SEK, particularly when ecological knowledge is documented in sources or forms unfamiliar to and marginalized by scientists. When an ancient O'Odham song describes the behavior of a hawk moth with rich and reliable detail, for example (Kimmerer 2002, 434), an unfamiliar audience might appreciate its aesthetic and cultural value yet fail to understand the song's discursive significance as an empirical report. More generally, the evidential significance of storytelling and oral histories are often overlooked by scientists and others who assume scientific speech acts such as data reporting and analysis must be delivered in a (quite specific) written form (Kimmerer 2002, 434; Whyte 2018b).

Hermeneutical Ignorance and Ecological Knowledge

Another TEK-related phenomenon of epistemic injustice involves settler colonialism toward misunderstood indigenous practices, where observers assume these practices to be nonsensical, superstitious, or otherwise irrational because these outside observers have failed to dispel their own hermeneutical ignorance. In these cases, the indigenous practices in question are genuinely meaningful, grounded in a rich, informed understanding of the relevant ecological relationships. But since outside observers are unfamiliar with these practices and relationships, they lack the conceptual resources to understand what's going on and the humility to realize they should learn more before drawing conclusions. Consider, for example, European explorers and immigrants to the northern Great Lakes, struck by the abundance of wild rice harvested by native peoples but also the seemingly inefficient way they did so. They were puzzled by the fact that "the savages stopped gathering long before all the rice was harvested," Kimmerer (2013, 181) explains. "The settlers took this as certain evidence of laziness and lack of industry on behalf of the heathens." Without understanding the ecological significance of leaving enough rice for other animals and for healthy harvests into the future, the newly arrived settlers could only see these sustainable gathering practices as wasteful, speaking to native peoples' questionable characters rather than their wisdom.

Willful hermeneutical ignorance also characterized the report on livestock safety issued by the state of New Mexico after the Church Rock nuclear spill. As we noted above, the conclusion that contaminated meat was safe for consumption was predicated on a glaring ignorance of traditional Navajo dietary practices. More generally, settler colonialism is essentially predicated on willful hermeneutical ignorance. As Whyte (2016a, 2016b) explains, settlers do not merely bring along their beliefs, values, and practices; they also impose these on their new home, undermining and driving out indigenous beliefs, values, and practices as though they had never existed, as though there *were* no significant beliefs, values, or practices there before settlers arrived. At its most complete, settler colonialism turns willful ignorance of indigenous knowledge into conventional wisdom, that this ignorance isn't really ignorance because there was nothing worth knowing. In this way, settler colonialism doesn't just fail to overcome ignorance of TEK, but actively constructs it.

Epistemic Objectification and Governance Injustices

Part of what makes TEK an interesting concept for a wide range of epistemic injustices is that the relationship between SEK and TEK is dynamic. More scientists have come to appreciate the relevance of TEK to their work. This doesn't mean that testimonial, participatory, and hermeneutical injustices are now entirely in the past; they continue, alongside other epistemic injustices now made possible by growing yet incomplete respect for indigenous peoples' epistemic agency. Epistemic objectification is perpetrated in that middle ground, where listeners recognize there is valuable information to be had while failing to recognize speakers as not just sources but knowers themselves. "So for example if a scientist treats Indigenous peoples as primarily interview subjects, that may completely ignore what the Indigenous peoples are trying to do in their own right," Whyte (2018b) says. Unlike earlier generations who might have dismissed TEK as distracting superstition, this sort of scientist genuinely sees its *supplemental value* ("the value of Indigenous knowledges as inputs for adding [i.e., supplementing] data that scientific methods do not normally track") though not its *governance-value* ("as irreplaceable sources of guidance for Indigenous resurgence and nation-building"). The issue here is not about scientists interviewing members of indigenous communities, but their seeing these communities and their members as interview subjects only, not potential collaborators or indeed independent inquirers. If indigenous knowers are recognized as epistemic subjects, not merely objects, crucial questions that are closed off by epistemic objectification are opened up. For example, if TEK has supplemental value for scientific knowledge-making, how can SEK have supplemental value for indigenous knowledge-making? If scientists must evaluate indigenous people as (un)trustworthy informants, how can scientists also view themselves as (un)trustworthy for indigenous people?[11] If TEK is understood as a socially ecologically situated way of knowing rather than as a body of propositional knowledge, can scientists engage with such ways of knowing without first building actual collaborative relationships with indigenous people who are so situated? Are scientists (and their funders) open to true collaboration rather than exploitation and appropriation?[12]

We should not assume that each instance of epistemic injustice involving TEK must fit into one of the categories identified so far; insisting as much may perpetrate further contributory injustice against already marginalized knowers. One notable phenomenon of epistemic injustice that seems to extend beyond these existing categories is what we might call *epistemic governance injustice*, building upon Whyte's distinction between supplemental- and governance-value of indigenous knowledge. In cases of epistemic governance injustice, I suggest, indigenous people are wronged in their capacities as knowers when they are unjustly prevented from developing and maintaining the ecological knowledges they need for planning, adaptation, and other governance purposes.[13] Consider, for example, the conflict between Karuk tribal practices and US policies on dams and fire management in California. Norgaard explains that Forest Service fire bans and river dams in the twentieth century undermined Karuk ecological knowledge and food systems (2014b, 73). As with other epistemic injustices, the wrongdoing perpetrated is political and epistemic: the problem here concerns tribal sovereignty, but more specifically the fact that good governance requires robust and ecologically responsive ways of knowing. As Reed puts it, "Criminalization of cultural practices matters for sovereignty because it directly prohibits the enactment of practices needed for the generation of knowledge" (Norgard 2014a, 22).[14]

Other epistemic injustices can undermine their victims' epistemic power through credibility misattributions, misunderstandings, misinterpretations, normatively inappropriate dissent, and doubt. For epistemic governance injustices, TEK can stagnate due to legal, political, and socio-economic impediments to sustaining ongoing knowledge. So understood, the wrong of epistemic governance injustice may remind some readers of Coady's (2010, 2017) analyses of epistemic injustice as distributive injustice, such that epistemic resources are distributed inequitably across social groups. Distributive epistemic injustice is relevant to ecological cases: when signs warning of the Rio Puerco's radioactive contamination were posted in English, for example, or when US regulatory agencies invested far more in understanding what went wrong in the nuclear disaster at Three Mile Island than the one at Church Rock in the same year (Brugge et al. 2007, 1598). What makes epistemic governance injustice distinctively different is that the problem concerns not how collective resources are distributed among groups but how these groups' knowledge-generative practices are permitted or prevented. In the case of Karuk ecological knowledge and food-systems management, the epistemic governance injustice is not that the US Forest Service failed to invest in salmon or huckleberry research, but that Forest Service policies undercut the tribe's own epistemic practices needed for ecological planning and governance.

TEK AND EPISTEMIC REPAIR

A review of the contemporary literature on epistemic injustice indicates that the most attention has been given to the following areas, in descending order of emphasis: phenomena of epistemic injustice and attendant harms; consequences and repercussions; individual or structural changes to prevent further injustices; and restitutive or reparative responses.[15] My aim is to urge greater attention to the last of these, in particular by taking a relational approach to epistemic repair. I do not mean to present reparative epistemic injustice as most important nor to denigrate work on epistemic injustice that focuses on other areas; phenomena of injustice must be well understood in order to attempt relational repair in their aftermath. In developing better epistemic practices, how can such practices be made meaningfully *reparative*: not only recognizing prospects for epistemic improvement going forward, but more specifically responding to experiences and perpetrations of epistemic injustice? With this question we are in the realm of nonideal theory, which has enjoyed considerable attention from ethicists and political philosophers but is also a familiar theme in epistemic injustice. We can see this emphasis in Fricker's *Epistemic Injustice* (2007), inspired by Judith Shklar's (1990) example in political theory not to take success as the presumptive object of analysis, as well as Jose Medina's *Epistemology of Resistance* (2013, 13), which follows Elizabeth Anderson (2010, 2012) (and John Dewey before her) in committing to a "methodological primacy of the non-ideal."

I suggest that an approach to relational repair that is extended and applied to social-epistemic relationships and communities is relevant to epistemic justice for two reasons. First, to the extent that epistemic injustices are epistemic *and* ethical wrongs, they are candidates for moral repair. Beyond this initial point, the harms caused by epistemic injustice also leave multiple parties in need of distinctly epistemic repair. Among the objects for epistemic repair are victims' epistemic agency, including their capacities for knowledge, self-trust, and good-informant status eroded by testimonial, hermeneutical, and other injustices. An epistemic relationship or an entire epistemic community may need epistemic repair, constituted as they are by trustful collaboration, division of cognitive labor, and reciprocal accountability (Walker

2014, 113).[16] A relational-reparative approach to epistemic justice should build on Emmalon Davis's recognition of the need to repair social-epistemic environmental conditions, not just our own epistemic and ethical characters, and repair not only the environments in which perpetrators and their victims are situated but also the communities in which we are members. We must attend to our epistemic communities and their constitutive relations of trust, recognition, and accountability.

Whatever else it is, epistemic repair must be a multi-agential process, not just something that perpetrators, victims, and third-party community members *realize*, but something we *do*. As with moral repair, a process of epistemic repair will involve perpetrator and community admissions of epistemic wrongdoing, apologies and amends in accordance with victims' subjectivities, renewed trustworthiness, and victim forgiveness. Let us consider how each of these elements of epistemic repair function in the aftermath of injustices concerning TEK.

Acknowledgment

Rectifying injustice requires not only reparative acts but also acknowledgment, specifically acknowledging one's actions as redress for one's wrongdoing. "Without that acknowledgement of error," Boxill (1972, 118) argues, "the injurer implies that the injured has been treated in a manner that befits him; he cannot feel that the injured party is his equal." Acknowledging one's responsibility for perpetrating an epistemic injustice is important for several reasons, the first of which concerns amelioration as iterative adjustment. A perpetrator who cannot or will not admit his epistemic wrongdoing to himself or others fails to learn, by reflecting on the experience itself or by taking up others' criticisms and recommendations in response. Absent acknowledgment of his wrongdoing, the perpetrator participates in the process of epistemic justice only abstractly, a sort of generalized project of epistemic improvement without attending to his specific complicity and positioning within it.

Another way in which acknowledgment is significant concerns what Heidi Grasswick calls *epistemic trust injustices*, which occur when, "due to the forces of oppression, the conditions required to ground one's trust in experts cannot be met for members of particular subordinated groups" (2017, 319). An acknowledgment of one's responsibility for perpetrating epistemic injustice does not complete the process of epistemic repair, nor on its own license renewed trust. It is a significant step nonetheless, without which victims of unacknowledged epistemic injustice otherwise are deprived of the considerable practical and epistemic benefits enabled by healthy trust relationships. For Kimmerer and others who see SEK and TEK as importantly interlinked, both of these ways of knowing benefit from cross-pollination, which is difficult if not impossible when the conditions for trust in scientific experts cannot be met for indigenous knowers.

Acknowledging the perpetration and experience of injustice is also important as reaffirmation of victims' community standing to demand accountability for wrongdoing. This is a reason why, when perpetrators fail to acknowledge their wrongdoing, recognition of that wrongdoing by the broader community plays a vital role. For epistemic reparative justice, such affirmation is no less important. Epistemic injustice erodes its victims' social-epistemic standing: as credible testifiers, discursive participants, collaborators (not just useful sources of information) and individualized knowers (not just token representatives). Perpetrators' and communities' failures to acknowledge victims' experiences as not just unfortunate but *unjust* leaves social-epistemic erosion unchecked and unattended. The longer this goes on, the clearer it becomes that victims of epistemic injustice are not treated as equal members of their relevant epistemic communities, no matter what generic assertions of epistemic equality might be made.

Amends

An account of epistemic repair that starts and ends with acknowledgment of injustice would be not only theoretically impoverished but practically inadequate for victims and perpetrators of epistemic injustice. We see this in the bewilderment and even indignation of perpetrators who assume that admitting their wrongdoing should be enough to restore an epistemically adequate status quo. What else is necessary? A retributive approach to epistemic justice might demand punishment; a restitutive approach might call for perpetrators to compensate their victims. Yet neither punishment nor compensation have clear analogues in epistemic contexts. One might try to interpret victims' or other parties' unwillingness to trust, consult, or otherwise collaborate with perpetrators of epistemic injustice as a kind of punishment. But I think that interpretation obscures how, in light of perpetrators' dismal track records, refusal simply may be prudentially and epistemically rational, regardless of one's retributive desires or lack thereof.

Alternatively, one might make a case for epistemic compensation in terms of giving credit previously withheld, filling hermeneutical gaps, reassessing misinterpreted speech acts, and so on. While each of these ameliorative measures might play a role in some processes of epistemic repair, it is not as compensation to balance the social-epistemic ledger. Insights and observations silenced by testimonial quieting or smothering can be communicated on another future occasion, perhaps between the same speaker and listener, given genuine acknowledgment of wrongdoing and subsequent relational repair. But to see this as compensation making up for earlier injustice would be to ignore the dynamic, iterative nature of epistemic agency and knowledge production. Filling a hermeneutical gap that led to hermeneutical injustice may be epistemically beneficial going forward, but it cannot repay victims of this injustice for the practical and epistemic harms they experienced in having some part of their experience obscured to social understanding in the interim. A compensatory approach to epistemic justice must be partial at best, and furthermore it risks perpetrating contributory injustice by representing compensation as canceling the harms of an injustice, contrary to victims' experiences of these injustices and their aftermath.

To argue against payback for epistemic injustice is not to insist acknowledging responsibility is enough for epistemic repair. We can understand perpetrators' efforts to make things right as attempts to make amends, where amends are not compensatory but rather communicative acts, underwriting processes of relational repair in alignment with victims' perspectives and priorities. Victims of epistemic injustice are being asked to trust in the aftermath of experiences that might constitute powerful evidence to the contrary: that their unjust treatments notwithstanding, they *will* be believed, included, understood, and so on. Amends enable this, which is why prioritizing victim subjectivities is so crucial for determining actions and outcomes needed to make amends after perpetration, experience, and acknowledgment of epistemic injustice. Those who commit epistemic injustices might think they know best how to show their contrition and trustworthiness, and yet an inward-oriented project of self-improvement, while perhaps worthwhile in other ways, misses what epistemic repair is supposed to address.

Prioritizing victims' perspectives in determining the appropriate amends for epistemic repair need not assume such victims' infallibility, magnanimity, or perfection otherwise construed any more than epistemic relationality assumes unconditional and absolute trust and trustworthiness. Making amends is itself a context of epistemic dependency for which perpetrators and victims may or may not be ready to participate. A listener may take responsibility for perpetrating an epistemic injustice against a speaker, for example, yet when it comes to making amends, does this listener actually trust this speaker to identify them? Do they understand amends as the speaker does; can they hear what the speaker is saying, not merely what they want to hear? Whether epistemic repair is even feasible or desirable is itself something different parties to the prior epistemic injustice may reasonably dispute.

Forgiveness

To the extent that epistemic injustice is also moral wrongdoing in need of repair, forgiveness has a crucial role. But is forgiveness also epistemically significant? If so, does it come in degrees, or is it an all-or-nothing threshold?

Perhaps you forgive me enough to enter into honest dialogue on some matter, but not enough to allow me to speak on your behalf. Recall Dotson's notion of testimonial smothering: perhaps you forgive me enough to testify to me on some things, but not on others. If forgiveness is part of a process by which relationships become morally adequate without assuming a morally adequate status quo, we might likewise understand forgiveness for epistemic injustice as enabling a kind of *epistemic adequacy*. Forgiveness so understood is a relational threshold, below which a range of cross-cultural collaborative projects are rendered impossible, and above which different kinds of epistemic relationships may be more or less trusting and trustworthy.

TEK AND SECOND-ORDER EPISTEMIC INJUSTICES

In the prior section I outlined a relational-reparative approach to ameliorating epistemic injustice, not merely in the senses of learning and improving on our past epistemic wrongdoings but acting to repair them. Extending reparative environmental justice into epistemic contexts does raise some difficult questions and complications, particularly concerning the potential for second-order epistemic injustices against TEK and knowers.

How can a process of epistemic repair itself go wrong? If some damaged relationships ought not be repaired at all, then as with moral repair, efforts at epistemic repair are inappropriate and unwelcome. Epistemic repair can also be harmful when one or more parties to repair perpetrates additional epistemic injustices in the process. Consider a situation in which a listener extends too little credibility to a speaker's testimony owing to some identity-prejudicial stereotype, and thus she perpetrates testimonial injustice; yet when this listener attempts to acknowledge her offense and make amends later, these reparative efforts are themselves misinterpreted, misunderstood, and summarily dismissed due to her own hermeneutical marginalization. How can a relational approach to epistemic repair make sense of this sort of interlocking, iterative epistemic injustice? Two explanations present themselves. One option is to continue to prioritize victims' subjectivity while recognizing the possibility of conflicting second-order epistemic injustices. Another option is to move to a conditional or defeasible priority for victim subjectivities, with certain necessary limiting conditions to be specified. The first of these seems to be the more promising direction. Given intersectional systems of power, privilege, and oppression, a perpetrator of one epistemic injustice might experience a second epistemic injustice in the aftermath of the first, particularly in those cases where the initial injustice remains still unresolved, relationships are in disrepair, and trust is in short supply. As Dotson's account of contributory injustice reminds us, we cannot overlook the very real possibility that the perpetrator of an initial injustice (or willfully ignorant bystanders) might commit further epistemic injustices in its aftermath, whether against the same victims wronged in the first place, active bystanders, or unfortunate third parties.

Consider, for example, the potentially fraught question of determining appropriate amends and conditions for forgiveness in the aftermath of epistemic injustices against TEK and knowers. Those who do not recognize their perpetrations of epistemic injustice *as* wrong are unlikely to give much thought to apologies, amends, or forgiveness. Yet those who do acknowledge and apologize for their epistemic wrongdoings might still push back against the ways in which amends and forgiveness are articulated. For such perpetrators, the conditions of forgiveness might seem too demanding, or the amends required seem too tangential to the wrong that they have done. Perhaps they really do not understand what their victims are asking of them, or they suspect that those they have wronged are ill-equipped or unable to really know what they need in order to rebuild trust and forgiveness. The irony is that these reasons for resistance might themselves reflect identity prejudices, willful hermeneutical ignorance, objectification, and other epistemic errors, and in rejecting or second-guessing what is required for amends or forgiveness in the aftermath of one injustice, they can cause further testimonial, participatory, hermeneutical, discursive, or other epistemic injustices.

The issue of contributory injustice raises the concern of whether the call for epistemic repair might itself perpetrate second-order epistemic injustices. Recall Berenstain's notion of epistemic exploitation: does a relational-reparative approach presume that oppressed persons must educate privileged persons on the nature of their epistemic oppression? We can anticipate how efforts at reparative justice might be tried in ways that epitomize Berenstain's critique. Yet perpetrators of epistemic injustice seeking to make amends and rebuild their trustworthiness can be guided by victims' perspectives and priorities without pushing victims to walk them through the reparative process. It is one thing for me to listen to you and another for me to compel you to speak to me, repeatedly and on demand. Indeed, when it comes to epistemic repair concerning TEK, oftentimes victims *already have* expressed themselves. Their testimonies are already available to be heard, even if now-penitent perpetrators were not previously listening or able to really understand or properly interpret what was being said.

This point about existing testimony as yet unrecognized by socially privileged listeners raises another possibility for epistemic injustice in this project, one of willful hermeneutical ignorance and contributory injustice. How might my discussion of epistemic repair itself ignore or serve to obscure existing indigenous knowledge on reparative justice, thereby contributing to the social-epistemic marginalization of indigenous knowledge and knowers?

This dismal possibility must be considered. I think that what warrants our attention here is the value of a relational-reparative approach to ecological epistemic justice, rather than any claim to novelty in the analysis I have offered here. To whatever extent my analysis proves convincing, then on methodological and ameliorative grounds, this should not sap but strengthen the imperative to seek out, recognize, and engage with marginalized knowledge on reparative justice.

NOTES

1. See also Millard et al. (1983).
2. For more on ignorance as a passive and active construct, see Proctor (2008, 6–8).
3. Furthermore, Paul said, "we want the lands and water and people and livestock who have been contaminated by the UNC spill decontaminated. We want our land, our people, our livestock, and our way of life restored as near as possible as it was before UNC and Kerr-McGee and their friends came to our land" (US House of Representatives 1979, 5).
4. Clifford's categorical thesis on the ethics of belief, that "it is wrong, always, everywhere and for anyone to believe anything on insufficient evidence," has since been subject to trenchant criticism by William James (1896), among others. For present purposes that wider claim is not needed, only the more limited lesson about irresponsible belief in cases like the shipowner and Church Rock.
5. See for example Berkes (1999) and Nakashima et al. (1993, 2012).
6. See for example LaDuke (1994), Pierotti and Wildcat (2000), McGregor (2008), and Kimmerer (2013). As McGregor (2004, 391) puts it, indigenous knowledge "is not just knowledge about relationships with Creation or the natural world; it is the relationship itself."
7. Whyte (2013, 10) further explains, "The significance of this point is that when the concept of TEK is used, it really points to the possibility that there are cross-cultural and cross-situational divides that make it so that non-indigenous parties cannot expect their own assumptions to apply to indigenous contexts."
8. On excess, Fricker says, "At a stretch, this could be a case of injustice as distributive unfairness—someone has got more than a fair share of his good—but that would be straining the idiom, for credibility is not a good that belongs with the distributive model of justice" (2007, 19).
9. See also Fricker (2006), Wardrobe (2015), and Kidd and Carel (2017).
10. Battiste and Henderson (2000, 36) identify the risk that TEK is assumed to be uniform across all indigenous peoples; see also Berkes (1999, 145–47) on TEK and the myth of the exotic other.
11. "From my perspective," Whyte (2018b) explains, "scientists first need to understand their own positions in relation to Indigenous peoples. . . . That is, when scientists, working for an institution, government agency, or university approach an Indigenous nation, the scientists represent them-selves as participating in the interests

of the U.S., a school, or the corporations who donated the research money. While the scientists themselves may not agree with the agendas or ideologies of the settler sovereigns or business interests, they are inextricably acting on their behalf *in some way* according to the perspectives of many Indigenous peoples."

12. On epistemic appropriation generally, see Davis (2018); on appropriation of TEK specifically, see Shiva and Holla-Bhar (1993), Shiva (1997), and Bannister and Solomon (2009).

13. To be clear, I am focused on a specific sort of epistemic governance injustice that I take to be particularly relevant to TEK and reparative environmental justice. More generally, we might say that an individual or group experiences an epistemic governance injustice when they are unjustly prevented from developing or maintaining knowledge for planning, adaptation, and other governance purposes.

14. See also McGregor (2004, 396). As Whyte (2018c) puts it, "As settler states are here to stay, they have instantiated and enforce laws, economic policies, and practices of cultural and political domination that leave Indigenous peoples with little space to plan both creatively and practically about what to do in the future."

15. For a thorough literature review on epistemic injustice, see McKinnon (2016).

16. See also Hardwig (1991) and Kitcher (1993).

Chapter 7

Reparative Environmental Justice in the Chicago Wilderness

Reparative justice provides tangible, practical ways not only to recognize our own and others' wrongdoings but to act with direction and purpose in response to that recognition. We need not do this alone. In his introduction to *Healing Natures, Restoring Relationships*, Robert L. France (2008, 8) observes that "community-based environmental restoration—the reparation of natural systems and the relationships of humans to them and to each other—is offered as a positive means by which to escape from the lonely apathy" common of many ecologically minded people. Here France invokes the distinctively relational approach to ecological restoration we can find in Bill Jordan, Robin Kimmerer, and Marion Hourdequin: as France puts it, "Environmental restoration is about restoring a personal relationship with the natural world, through both conscious thought and physical action" (2008, 6).[1] Jordan, in particular, has emphasized how that distinctive feature of restoration as mindful physical activity transforms it into a meaningful ritual. As he argued in *Restoration & Management Notes* throughout the 1990s, it is a ritual we perform not in isolation but in relation: "Restoration projects tend to become community events. They carry participants not away from society, but back into it, and through it, back into the natural community" (1992, 112). As previously discussed in chapter 3, I find this sort of relational account of environmental restoration compelling, and yet complications linger. Specifically, how are we to make sense of real-world situations in which restoration seems to strengthen some of our relationships while exacerbating a sense of disconnection and distance for others? It is with this challenge in mind that I turn to the Chicago Wilderness.

ORIGINS OF THE CHICAGO WILDERNESS

The name "Chicago Wilderness" refers both to a multi-organizational alliance and a geographical region, to both people and a place, which fits with Liam Heneghan and his coauthors' (2012, 85) characterization of Chicago Wilderness as a social-ecological system. The name is meant to be a bit cheeky: not ironic exactly, but more aspirational. The conflicting connotations are baked in: *Building. Wilderness. In a city.* The Chicago Wilderness is not constrained to city limits, to be sure: the region extends from southeastern Wisconsin through Chicagoland into northern Indiana and the southwestern corner of Michigan. It includes public land in the Forest Preserves of Cook County (FPCC), the Chicago Park District, parks and preserves in "collar counties" (Lake, Will, DuPage) around Chicago, and other government, nonprofit, and private holdings across the four states, comprising a variety of biomes and ecotones including prairies, woodlands, wetlands, lakes and rivers (Chicago Wilderness 1999, 2016).

What I find affecting and effective about the name is that it brings to the surface underlying assumptions for wilderness, including aesthetic assumptions that wilderness should be sweeping, grand, quietly beautiful. That is not the City of Big Shoulders. The allure of Chicago Wilderness is not that it is pristine. Neither is the intention to paint Chicago as a jungle, on the classical view of wilderness as a desolate place unfit for human civilization.[2] This raises a conceptual question: what makes something wilderness? If wilderness is defined as wherever humans do not actively and deliberately manipulate their environment, as Dwight Berry (1998, 125) puts it, how could Chicago be such a place?[3] How could it be, especially when the mission is not to leave things alone but to pursue dynamic restoration projects? "Chicago Wilderness" is not apologetic, but celebratory: wilderness as biodiversity, as resilient ecological community, and a celebration that such things are possible not only in the Alaskan bush or the Galapagos Islands but also *here*. The name, Heneghan et al. (2019, 3) write, "crystallizes the incongruity of having globally important habitats embedded in a major urban area."

Many readers may be distressed but not especially surprised to learn that nearly all of the vast prairie that stretched across the Prairie State for millennia is gone. "Of the estimated 8.9 million hectares of prairie originally in Illinois, 93 hectares remain—a decline of 99.9%" (Heneghan et al. 2012, 77; see also Buckley 2020). Perhaps more surprising is how much of the remnant prairie acreage is within the Chicago metropolitan area: as Heneghan et al. (2009, 65) explain, "Many of the higher quality residual fragments of presettlement habitat are preserved in the hinterland of the city, whereas, by virtue of the fertility of its prairie soils the extensive rural landscapes of the state are dominated by intensive agriculture with consequent depletion of biodiversity."

While grasslands and other ecosystems were displaced throughout much of the state in the nineteenth and twentieth centuries, botanist Jens Jensen, architect Dwight Perkins, and other Chicago community leaders advocated, and voters approved, the establishment of the nation's first forest preserve district in 1914. Then as now, its mission was delineated as follows:

> To acquire, restore and manage lands for the purpose of protecting and preserving public open space with its natural wonders, significant prairies, forests, wetlands, rivers, streams, and other landscapes with all of its associated wildlife, in a natural state for the education, pleasure and recreation of the public now and in the future. (FPCC 2020)

While this same mission has remained in place, it has been interpreted quite differently over the past hundred years.

Consider the tension between prairie and forest conservation, beginning with the curious choice to establish a forest preserve in a place where, Jensen himself said, the "predominating character of the landscape is that of prairie."[4] Natalie Bump Vena (2013) argues that the FPCC naming question is more than a harmless curiosity: "Cook County's prairies nearly disappeared during the twentieth century as District staff constructed a forest landscape to match the enabling statue's name and language."[5] From the 1910s into the 1960s, county commissioners did indeed acquire woodlands but also farmlands, wetlands, and prairies, with extensive tree plantings on the latter toward the goal of reforestation, "thus creating the illusion that they were establishing something that had previously existed on the land" (Vena 2013). In an attempt to protect existing and newly planted trees, FPCC then adopted fire-suppression policies, preventing wildfires and discontinuing the traditional land-management practice of setting prairie fires conducted by the Potawatomi and other tribes.[6] "Where buckthorn, hawthorn, foreign honeysuckle, dogwood, and introduced trees did not take over," says William Stevens (1995, 38), "the Forest Preserve District routinely mowed open areas that once would have been a thick carpet of prairie and savanna grasses and flowers."[7]

Laurel Ross, longtime restorationist and chair of the Chicago Wilderness Executive Council, describes the ecological challenge for resource management and conservation thusly:

> As is true in so many places, these previous landscapes have been under siege. Decades of fire suppression and invasion on a massive scale by aggressive species [. . .] have caused serious degradation, with remnants of rare communities shrinking, and species slowing dropping out of communities where they have flourished for millennia. . . . But ultimately it is sheer neglect, and the absence of

support for the hard and expensive work that must be done to reverse ecological degradation that are the worst consequences of public apathy. (Ross 1997, 19)

Thus the driving motivation for restorationists like her, Ross explains, "was the realization that there existed a world-class biological resource in our region, but that its well-being—its very survival—was seriously threatened" (ibid.).

A renewed appreciation for the value of tallgrass prairie and the need to maintain and restore it began in the 1960s, with fruitful collaborations among Floyd Swink and Ray Schulenberg (at the Morton Arboretum), Bob Betz (an ecologist at Northeastern Illinois University), and David Blenz and Chuck Westcott (directors of FPCC's Camp Sagawau and Little Red Schoolhouse, respectively) (Jordan 1987; Vena 2014). Throughout that decade, Swink taught naturalists and others about prairies and their unique plants; his 1969 book *Plants of the Chicago Region* and its subsequent editions (Swink and Wilhelm 1994; Wilhelm and Rericha 2017) continue to be primary references for restoration ecologists. Westcott, Betz, and Schulenberg explored remnant prairies all throughout the region: in vacant lots, cemeteries, along railroad tracks. Blenz sowed a small prairie at Sagawau in 1965, followed soon after by similar projects at Little Red Schoolhouse, Crabtree, and Sand Ridge Nature Centers (Vena 2014). When construction began on Fermilab in Batavia in the mid-1970s, Betz and Swink served as consultants on its massive prairie restoration (Betz 1986; Betz et al. 1996). As Betz (1972) wrote in his introduction to *The Prairie: Swell and Swale*, "with hard work and luck (such as sufficient rainfall at the time of planning) fairly good prairies can be established within a few years on suitable land."[8]

Betz's remarks were inspirational to Steve Packard, then a former antiwar activist with what he called "an empty hole in my life" (Stevens 1994, 47). Stevens explains his reaction thusly:

> Taking a leaf from his activist days, Packard decided that what the neglected prairie remnants needed was a constituency. He thought "I can bring people up here and show them." And then one day it dawned on him that, no, if the prairies were to survive, they needed congregations, as he started calling them, of people who could interact with the land—who could develop an emotional bond with it. The way to achieve that, he thought, was to get people to come out and work on the land and help restore it to health. (Stevens 1994, 50)

While prairie restoration was already underway at Fermilab and elsewhere, defenders and critics alike would agree with Christopher Johnson's (2019) judgment that "it was Packard who turned restoration into a community project." What Packard started with twelve Sierra Club members gathering seed near Somme Prairie on August 6, 1977, has grown in its size and scope over the years: in 2017 nearly 20,000 volunteers dedicated more than 60,000 hours cutting, seeding, and contributing in other ways to FPCC restoration projects, not to mention contributions to similar projects in other preserves and parks (Johnson 2019; Chicago Wilderness 2019; North Branch Restoration Project 2020).

A celebration of human participation in ecology pervades Stevens's book *Miracle under the Oaks*, which offered a nonspecialist's account of prairie and savanna restoration undertaken by Packard and his team along the North Branch of the Chicago River. Stevens celebrated Packard and his fellow North Branchers, yet his book also served as a sort of Bible for some of Packard's fiercest detractors such as Mary Lou Quinn and others from the anti-restoration group Trees for Life (Joravsky 1996; Friederici 2006, 117). As Alf Siewers (1997, 25) and Debra Shore (1998, 11) both suggest in their reflections on the Chicago restoration movement, the publication of *Miracles under the Oaks* was in a way an unintentional set up for a fall. It brought attention to restoration work that had been done quietly through trial and error for years, offering readers a warts-and-all gloss on the process.

After that, things moved quickly. *Miracle under the Oaks* was published in 1995, Chicago Wilderness as an organization was founded in 1996, and within the year Cook County Board President John Stroger issued a moratorium halting restoration activities in the Forest Preserves (Fegelman 1996; Kendall 1996; Van Matre 1996). As Packard noted (2015) twenty years later, "Chicago Wilderness would for a time come to be a respected world model of conservation partnership. It deserves credit for many things. On the other hand, it was not well equipped to handle its biggest challenge, which arrived just one month after its auspicious kick-off."

RESTORATION AND ITS DISCONTENTS

Opposition to ecological restoration in the Chicago Wilderness in the 1990s took several forms, to varying degrees of social impact. Let's consider these grounds for opposition one by one.

Scientific Opposition

Packard's theories and practices stood in contentious relationship with conventional wisdom among ecologists, sometimes productively so. Like many other participants in the North Branch project, Packard had no formal training in ecology, botany, or biology, yet few scientists could rival his experience in the field.[9] When FPCC brought Betz to assess the North Branchers' work in July 1978, Packard was relieved and encouraged to hear Betz describe what they had created as incipient prairies. "Betz's imprimatur was critical," Stevens (1994, 59) explains: "It gave the North Branchers a credibility they had never had up to that point. And largely on the basis of his say-so the Forest Preserve officials decided that the North Branch sites were to be managed as prairies and gave the restorationists formal permission to sow seeds and clear brush and trees." For his part Betz occupied a middle ground between the volunteer restorationists and scientific establishment. Consider for example the use of fire. Seeding and clearing brush were not enough, Betz and Schulenberg argued. They knew that the Potawatomi and other tribes had long used fire for prairie management, and so they promoted and practiced controlled burns in conservation and restoration. Betz experienced little direct opposition on burning from other ecologists early on, he says, but as prairie preservation and restoration grew in popularity and government agencies and conservation organizations sought the advice of scientists, they invariably counseled against it (Stevens 1994, 118–19).[10] Fire become a point of contention among FPCC officials as well. Now working together, Packard and Betz sometimes clashed with longtime forester Sam Gabriel, who had spent his career planting trees throughout the preserves and saw burns as contrary to his mission. They compromised, with Packard and Betz proceeding with controlled burns at several sites and Gabriel securing an exemption for Sauganash Prairie Grove and the white oaks there (Stevens 1994, 68).

Though Packard had been frustrated that Gabriel had not let them remove the oaks to restore uninterrupted grassland, he began to think that perhaps Gabriel had a point after all. "The oaks and the trilliums under them were part of the holy thing we were groping for" (Stevens 1994, 81). Packard's revised approach was not to let things be, nor to pursue reforestation, but "to eliminate the buckthorn that choked the oak groves on their site so that whatever natural relationship had once existed between the oaks and the prairie could be reestablished" (Stevens 1994, 82; Packard 1988, 14). He found encouragement in John Curtis's definition of savanna: neither totally prairie nor totally woodlands, but rather a grassland with trees, with up to half the area covered by forest canopy. Curtis framed the 50-percent coverage criterion not as a strict delineation between forest and savanna, but more of a convenient place to draw a useful distinction (Stevens 1994, 87–8).[11]

Packard was galvanized by Curtis's work and its relevance to the North Branch project, as he explains in his 1988 paper "Just a Few Oddball Species: Restoration and the Rediscovery of the Tallgrass Savanna."[12] He starts with an interesting failure: after years of weeding, burning, and seeding, Packard and his fellow North Branch restorationists found that prairies grasses had not taken very well in certain areas, particularly those closest to the trees. After some frustration, they noticed grasses thriving at Miami

Woods, another restoration site; these grasses were not what they had sowed, Packard observed, nor were they familiar to him as native prairie grasses. "Looking back, I realize that part of our problem was that we were thinking too much about prairie and weren't picking up on what this other community—the savanna—was trying to tell us" (Packard 1988, 17). He eventually identified the mysterious Miami Woods grasses in *Plants of the Chicago Region* (Swink 1969). They were indeed native species. Packard began compiling a long list of what he theorized to be distinctive savanna species, which he presented at the 1985 Northern Illinois Prairie Workshop and which was initially greeted, he says, with a skeptical yet genuinely interested scholarly reception (1988, 18).[13]

In 1992, Jon Mendelson and coauthors published "Carving Up the Woods," arguing against savanna restoration. Their critical aims were both specific and broad, with particular scrutiny on FPCC restoration goals and methods at Cap Sauers Holding and a call for concern on ecological restoration generally:

> Restorations are forever subject to the limitations of our understanding and to the imposition of our values. We see this increasingly in the destruction of species deemed, for one reason or another, "unsuitable" to a particular restoration. Subject inevitably to these constraints, all restoration plans and projects should be carefully evaluated against the alternative of letting natural processes continue, whatever their direction, without human interference. (Mendelson et al. 1992, 127)

Mendelson's specific criticism was that because savanna is not a unique biome, but rather "no more than a name applied to an arbitrarily defined segment of a particular continuum," it was therefore an improper management goal. "At Cap Sauer's Holding, and we suspect other sites, managers have chosen inappropriate landscapes to restore to savanna and open woodlands," he objected. "These choices force them to use inappropriate techniques to accomplish their goals" (1992, 128).

Restoration & Management Notes published several critical responses to Mendelson's paper, including letters from Richard Hyerczyk and Debra Lynn Petro, stewards at Cap Sauers Holding. "Those of us who do restoration at Cap Sauers and at similar sites in northeastern Illinois are not the ones who are carving up the woods," Petro (1993) asserted. "We are attempting to heal past abuses, to make the ecosystem whole again."[14] In his own response, Packard (1993, 10) said that Mendelson's argument "seems to be that the savanna's unusually 'transitional' nature sets it off from more legitimate communities like tallgrass prairies and maple forest."[15] Packard insisted not that tallgrass savanna must be a discrete entity rather than an ecotone between prairie and forest, but that the savanna is worth restoring either way. "The fact that a community is part of a continuum does not diminish its significance," he argued. "Most if not all communities are part of continua" (ibid.).

Animal-Welfare Opposition

Restoration activities in the FPCC and Chicago Wilderness widely generally focus on plants, soil, and water, but this does not mean animals are entirely unaffected. On the one hand, habitat restoration can have tangible benefits for both migratory and resident animal populations; on the other hand, some birders worry that tree and brush removal destroys existing habitat, and both restorationists and their critics recognize that some animals do perish in controlled burns.[16] But the opposition to restoration by animal activists that received the most public attention before the moratorium concerned the overpopulation of white-tailed deer. Unlike buckthorn and honeysuckle, white-tailed deer are native to the Chicago Wilderness region. Their population was believed to have grown to unsustainable level for two reasons: previous human interventions had eliminated their natural predators, notably grey wolves, and winters had been unusually mild in the late 1980s and early 1990s. White-tailed deer populations were a concern for the health and well-being of the deer themselves, but also for ecological conservation and restoration. Deer play an important ecological role in prairies, savannas, and woodlands, but in such large numbers and with few predators they were overgrazing wildflowers and other native flora, particularly in restored areas that had not had enough time to reestablish themselves. For this reason the North Branch restorationists supported Forest Preserve policies and practices of deer control, and in this way concerns about deer and concerns about restoration became linked in public debates and discussions.[17]

Yet animal activists stood in complicated relationship to the moratorium. Debra Shore (1997) reports that protests over deer population control voiced at public hearings (first in Lake County, then Will County, and finally Cook County) indeed drew critical attention to restoration projects in the Chicago Wilderness in the 1990s. But there are at least two complicating factors here. The first is that Packard, Ross, and their fellow restorationists were sympathetic to animal activists' concerns and made concerted efforts to find common ground. Consider this excerpt from Ross's "Thoughts on the Deer Problem," published in February 1992 in the North Branch Prairie Project newsletter *Prairie Projections*:

> Deer are beautiful and charismatic creatures. The prospect of shooting them is distasteful to many people, and the public debate on this subject has at times been emotional . . . Friends. This is not a dispute between deer lovers and deer haters. Let us keep the focus on the ecosystem. The inescapable fact is that

there is an acute and growing problem which will not improve until some of the animals are removed. (Ross 1992, 10)

For their part, many animal activists were not unsympathetic to restorationist concerns either. As Stevens explains, Cynthia Gehrie and other early critics of deer-control policies and practices in the preserves met with Packard and came away with some amount of reassurance and progress toward reconciliation. Gehrie was no less concerned about the white-tailed deer than before, but she was encouraged that Packard did not take the issue lightly, and she could appreciate that overpopulation was bad for the deer themselves and the larger ecosystem. Without renouncing the moral weight of killing deer, Gehrie came to accept and defend it as a necessary temporary measure for purposes of population control.[18]

Not all animal activists were as convinced as Gehrie, of course, and concerns about culling deer persisted. Another complicating factor is that while deer-welfare concerns brought greater public scrutiny to ecological restoration in the preserves, little if any of this greater scrutiny built upon these concerns. Mary Lou Quinn and others in Trees for Life did not cite deer population control as cause for alarm, nor did *Chicago Sun-Times* columnist Ray Coffey include this among his populist and procedural grievances against FPCC officials and restorationists. Most notably, the moratoria in Will County and Cook County prohibited controlled burns and tree removal for restoration purposes, but not tree removal for trail construction and maintenance nor shooting deer for population control (Fegelman 1996; Van Matre 1996).

Arborist Opposition

Perhaps the most frequently cited cause for concern about restoration in the Chicago region, both before and after the moratorium, was the removal of trees to create prairies and savannas.[19] This line of criticism included both a weaker version (trees are good, there is no need to remove them) and a stronger one (trees are actually relevantly better than prairie). Arborist opposition to restoration was often though not always connected to neighborhood-based opposition, including concerns about car noise, privacy, and smoke from nearby burns (Shore 1997, 30). "What was useless brush to restorationists was important buffer for Quinn, blocking the woods from the noise and traffic of the industrial world outside," wrote *Chicago Reader* columnist Ben Joravsky (1996). As Quinn put it to him, "These trees, even the buckthorn, are our buffer. Who are they to destroy that?"

Petra Blix has been another critic of tree removal for ecological restoration in the Chicago Wilderness. "I'm for restoration of appropriate areas, but I'm against cutting down forests to make something else. We don't have much forest left and I think we should protect what we have," she told Joravsky. "Only they say they're going to cut down the forests to save the oaks" (Joravsky 1996). This sentiment was not unique to Quinn and Blix. In a survey of Cook County residents soon after the moratorium, Barro and Bright (1998) found that residents were generally in favor of restoration, but opposed the use of herbicides and cutting mature trees.[20]

Years later, after the moratorium was lifted and restoration resumed throughout the preserves, Quinn explained her position thusly:

> My own heart of hearts is I wish we could get back to the 1960s and 1970s when it was shady and beautiful. It didn't get bad until man decided it was going to be a prairie. I'd like to have the prairies be nurtured back to what they could be. Plant some trees. Our area, with all these cars, we need trees. . . . A prairie is *one* thing—a prairie. It's *trees* that purify the air. (Friederici 2006, 123)

For her part Petra Blix (2007) also continued advocating tree preservation in restoration projects long after the moratorium ended.[21]

Procedural/Populist Opposition

Restoration opposition received a major boost in publicity from a May 12, 1996, piece in the *Chicago Sun-Times* with the headline "Half-Million Trees Face Ax" (Coffey 1996a). This was the first of many *Sun-Times* columns from 1996 to 1998 by Ray Coffey directing suspicion and skepticism against FPCC, Packard, and restoration in general. Coffey portrayed FPCC officials as oblivious and evasive and Packard and the North Branchers as unelected outlaws not playing straight with the public.[22] In a May 31 editorial, Coffey (1996b) drew his readers' attention to excerpts from *Miracles under the Oaks* as damning evidence of Packard's untrustworthiness.[23] The following passage was particularly notorious:

> From day one, Packard and his volunteers took pains not to attract the opposition of people whose sensibilities about nature, however uninformed, might be offended. "A very big part of what we did," says Packard, "was to be perhaps boringly cautious about things. We went to the most extreme lengths so that the Forest Preserve District would not have to deal with complaints about what we were doing." . . . So the North Branchers tred lightly, even sneakily. Their brush-clearing and planting operations took place as discreetly as possible. As a screen for their activities, they would leave a wall of brush in place along the bicycle trail or roadside. Behind the screen, they would cut a little brush here, a little brush there, and make their plantings. Eventually the plantings would

flower and everything would look stable and lovely, and only then would they cut away the roadside screens. After the vegetation had grown up, they removed unwanted invading trees (always called brush by the North Branchers, to deflect criticism) by girdling them below the vegetation line, where the killing cut could not be seen. Then the tree would slowly fade away and no one would notice. The whole idea was to keep people from becoming upset about destroying "nature" so that nature could actually be restored. (Stevens 1994, 66–67)

Careful readers of *Miracle under the Oaks* might reasonably find this passage not as damning as Coffey positioned it. For one thing, it might be exceptional rather than properly representative of Packard's guiding ethos.[24] Taken on their own terms the North Branchers' activities as described here might seem more like neighborly courtesy then trickery. To those who trust restorationists, the decision to keep work in progress out of sight was understandable. But for critics like Coffey and Quinn it was indicative of restorationists' less-than-transparent approach to working public land.[25]

Shore (1997), Packard (2016), Jordan and Lubick (2011, 186), and Ross all point to Coffey's steady stream of anti-restoration editorials in the *Sun-Times* as having a significant influence, socially and politically speaking, though for his part journalist Woodworth (2013, 115) sees this as a case of shooting the messenger. What is clear is that Coffey went past scientific and arborist opposition to advance procedural and populist objections to restoration, before the moratorium and after it had been partially lifted. Between November 1996 and January 1998, Coffey covered restoration once, then penned a dozen pieces on the subject until his retirement the next year.

AFTER THE MORATORIUM

Restoration stoppages in Cook and DuPage counties were lifted for most sites within two years, though a burn ban continued at some sites for a decade (Ciokajlo 2006). As Packard (2016) said, "bit by bit, stewards resumed cutting brush, then cutting invasive trees, then burning brush piles, and after a few years the program was pretty much as before, but wiser." Superintendent Steve Bylina promoted John McCabe to supervise FPCC restoration, which has continued productively for years with key contributions from employees, restoration technicians, participants in the Site Steward and Mighty Acorn programs, and other volunteers (Friederici 2006, 121). It is hard to escape the conclusion that renewed restoration work following the moratorium succeeded not by repairing damaged relationships with fierce critics (Coffey retired, and Quinn and Blix continued to criticize tree removal and ecological restoration for years after the moratorium) but by keeping a lower public profile and institutionalizing restoration more than before. In what follows, I want to draw attention to some significant ways in which Chicago Wilderness leaders have done—and going forward, can continue to do—more than just that.

In "Urban and Ecological Planning in Chicago," Margarita Alario distilled two main lessons from what became known by restoration practitioners and scholars as the Chicago controversy:

> First, "green" motives may conflict with each other, which can be seen clearly not only in the intense debate between conservationists and restorationists, but also in the conflicts between proponents of recreational, educational, and other uses. Secondly, efforts have been made to resolve these conflicts and avoid others, so restoration projects are currently more tentative, partial and sensitive to local input, and more information and publicity have been provided for education and debate. (Alario 2000, 490)

Let's build upon Alario's insights with some related lessons from the discussions of reparative environmental justice in the preceding chapters.

Identifying Our Restorative Aims

What does ecological restoration aim to restore, and why is it worthwhile? For Packard and the North Branchers in their painstaking, fulfilling work in the years documented in *Miracle under the Oaks*, their restorative aims were framed in primarily historical terms. The goal was to return restored lands to their presettlement conditions. There was nothing uniquely special about the 1830s, Packard argued. The 1839 Public Land Survey (PLS) is a snapshot of northern Illinois at the historical moment after the Treaty of Chicago, Black Hawk's defeat, the forced migration of the Potawatomi west of the Mississippi, and chartering of the city of Chicago (Heneghan et al. 2013, 340). There could be any number of equally valid reference points, Packard explained:

> The reason we look to the PLS notes is that typically they are the best single source of information on what the relatively stable, locally-adapted ecosystem was like. Taking the land back to 1738 or 1492 or 2,000 BC would serve just as well—if we knew how. As David Brower puts it, we are in the position of lost hikers trying to relocate the last spot at which they still knew where they were; once we get back on track, the idea is not to stand still but to proceed. (Packard 1993, 14; see also Woodworth 2013, 110–11)

It doesn't matter where we get back on track. Note that *on track* is understood in both historical and normative terms. If being on track represents the historical record of the changing ecology of a place, then the lands along the North Branch of the Chicago River and throughout the Chicago Wilderness are necessarily always already on track; deviations are impossible, no matter how degraded they might become. *On track* as Packard (following Brower) envisions it represents not so much what a place was before but what it should have been and should be, for some sense of *should*. So 1839 or any other presettlement reference point matters not in its own right but as an indirect indicator of ecological health, on the underlying assumption that these were times when ecological configurations were as they should have been—unlike, say, in 1939 or 1969.

Problems arise when we conflate the indirectly significant heuristic with the goal itself, take the heuristic for granted, or fail to sufficiently convey to other interested but as yet unconvinced parties why the heuristic is worthwhile. What comes across consistently in newspapers' and first-personal accounts of the Chicago controversy was that restorationists were working to restore sites to their presettlement conditions; not so consistently conveyed was why this was socially and ecologically valuable. "I listened to them but it didn't make sense," Quinn told Joravsky (1996). "Why would you want to turn back the clock to 19th-Century Illinois?" Another critic argued that "to select 1830 as the target for a turn to the past is without reason and completely arbitrary" (as quoted in Coffey 1996a). In his September 24, 1996, column on restoration public hearings, Coffey (1996f) wrote, "A good question for the County Board members to start with might be this: 'What relevance does the prairie restorationists' fixation on pre-settlement 1830s Illinois have to do with the reality of 1996 and a population of more than 5 million people, plus more than 3 million cars and trucks polluting the air around here?'"

DuPage County Commissioner Alice Peterson defended the restorationists' goal to "keep at least these acres the way they were" in pre-settler times so that "future generations can see how Illinois was before settlers came" (as quoted in Coffey 1996a). This answer might work for an arboretum, where the goal is about demonstration and education. But it misses and misrepresents the ecological value of restoration in specific biotic contexts, where ecosystem recovery makes a tangible, material difference for the humans, animals, plants, and microorganisms that make up the biotic community.

After the moratorium there was a notable shift in how the aims of restoration were identified. "Although an understanding of pre-settlement system has been an important guide to management," Heneghan et al. (2012, 81) observe, "attempting to recreate a faithful replica of the former state is no longer regarded by land managers as either a logistically or ecologically realistic goal." In contrast to Peterson's earlier appeal to restoration as historical education, when the burn ban was lifted executive director of the Nature Conservancy in Illinois Bruce Boyd said, "If we want to maintain our native biodiversity, we have an obligation to maintain these lands" (in Ciokajlo 2006). And by 2010 Packard described his goals for restoration in rather different terms than before, with Somme Prairie Grove as his illustrating example:

> The prairie and savanna, it is an injured thing, we are just trying to heal it and allow it to spread and increase . . . *if someone came to me and said this was a forest in 1830 I would say I don't care* . . . biodiversity is a precious and wonderful thing, the future will appreciate that we saved or resurrected some of it, and if under modern conditions the [biological] community behaves in a different way, I don't much care, that's all right. (Woodworth 2013, 111, emphasis in original)[26]

This shift from historical recreation to healing and restoring biodiversity aligns with the Chicago Wilderness Biodiversity Recovery Plan (1999) and a similar shift in the Society for Ecological Restoration (SER) definition of ecological restoration.[27] Previously put in terms of returning an ecosystem to its initial state, SER (2020) now defines ecological restoration as "the process of assisting the recovery of an ecosystem that has been degraded, damaged, or destroyed."

As Heneghan et al. (2012, 81) note, this new definition "does not refer to faithful reproduction of historical conditions but implies a notion of ecological 'health' which is translated in the Chicago Wilderness alliance as a suite of strategies that focus largely on the conservation of biodiversity in the region."[28] Understanding restoration in terms of assisting ecosystem recovery allows for an active role for nonhuman animals, plants, and microorganisms. Moving from an emphasis on historical recreation to fostering biodiversity also promotes a richer analysis of what successful restorative processes involve, and ecological restoration that is more responsive to the shifting demands of climate change.[29] Last but not least, this shift invites restoration practitioners to bring their work into closer conversation with scientific researchers, which many commenters see as a good thing for both the field of ecology and community-based restoration projects in the Chicago Wilderness.[30]

To be sure, some critics are not much more receptive to restoration for biodiversity than for historical authenticity. At the height of the controversy, Ben Joravsky facilitated a meeting between Petra Blix and FPCC site steward John Balaban. "They tried to be cordial, though they agreed on almost nothing," Jorvsky said. "He talked about biodiversity and she talked about trees" (1996). Trees for Life rejected restoration in terms of fostering biodiversity: "The controversy . . . is not about bio-diversity or restoration of degraded ecosystems. It *is* about *cutting down healthy trees* to create prairie or savanna where there is now forest" (in Shore 1997, 31, emphasis original.) How we identify the goals of restoration makes a difference, but this does not ensure all parties will agree on its value. Chicago Wilderness initiatives have not achieved a community-wide consensus, but they have been more successful than earlier restoration projects in articulating *why* ecological restoration is worth pursuing.

Ambiguities of Community

When Dave Egan succeeded Bill Jordan as the editor of *Ecological Restoration*, he began by explaining how his vision differed from that of his predecessor, in particular "what I regard as the exclusionary nature of Bill's interpretation of community" (Egan 2001, 67, 2004). Consider Jordan's assessment of restorationists and their critics after the Chicago controversy:

> Here critics of restoration are claiming their rights as citizens to equal treatment with respect to public land. More deeply, perhaps, they have in mind a conception of nature itself as a birthright, an idea of nature as free—free for the taking, so to speak—that reflects the experience of the frontier. The restorationists, on the other hand, are insisting on a very different idea. Having discovered restoration as a way of achieving membership in the land community, they are moving enthusiastically toward the creation of communities of a sort. I believe the restorationists are on the right track. (Jordan 2001, 1–2)

For both Jordan and Packard, part of what is powerful about ecological restoration is building a relationship with one's fellow restorationists and the land itself. "I remember this tremendous sense of confirmation," Packard said of his early work on the North Branch. "It was as if the ecosystem patted me on the back and said, 'Okay, kid, you've done well.' I had a very strong sense that we were *there*, that this was *it*" (Stevens 1994, 136).

This otherwise rich and rewarding sense of human-ecological community seems to exclude, or at least overlook nonparticipants and restoration critics as community members. It is not that the North Branchers or other restorationists were stingy with granting membership: as we have seen, one of the greatest achievements of restoration in the Chicago Wilderness has been the active participation of all sorts of people, not just experts but novices too, not just paid professionals but volunteers too. Certainly there has been conflict and disagreement within this motley community of restorationists, including conflicts over goals and methods and conflicts over different visions for ecosystem recovery and how ecosystems respond to human efforts to foster these visions for recovery. In that sense Packard and Jordan might agree with Egan's (2011, 67) conception of community "as a diverse collection of people who live in proximity to one another by choice or by fate, whose views on a variety of things may be dissimilar." But Egan's point is that if we take community repair seriously as a part of ecological restoration, we cannot ignore that our neighbors who are neutral or wary toward restoration are members of the community too.

Kimberly Phalen (2009) contrasts the approach to community involvement taken in the Forest Preserves with a similar Chicago restoration project at Montrose Point on Lake Michigan. Phalen argues that the Montrose Point project engendered less public opposition because birders, nearby residents, and other interested community members were invited to voice their preferences and priorities in the project design phase, and this community input was allowed to shape the project in concrete ways. For example, many birders opposed the proposed removal of a large growth of honeysuckle known as the Magic Hedge, which served as a stopping place or home for hundreds of species of birds (Phalen 2009, 183).[31] Honeysuckle is invasive in this region, and though it is not easily eradicated, many restoration projects target it to facilitate native-species recovery. For their part, historic preservationists sought to remove the hedge to better reflect Alfred Caldwell's original 1938 park design. Phalen describes how the Chicago Park District built these priorities into a vegetation plan with points of convergence and compromise among historical, ecological, and birding goals.[32]

As visitors to the Montrose Point Bird Sanctuary today will notice, the Magic Hedge remains. Is this a good thing? I argued that a strength of reparative environmental justice is its emphasis on the needs and perspectives of the victims of environmental wrongdoing in guiding practices of ecological repair. Among other things, this means for example restoring a tallgrass prairie not as restitution because that is how it had been before, nor as the cheapest available retrofit for derelict government or private lands, nor as a general public good because many people like (or don't like) prairies. If we approach habitat restoration as the work of ecological relational repair, the reparative question is not how keeping, removing, or trimming the Magic Hedge can balance stakeholder preferences, but what role the Magic Hedge may or may not play in making amends to human and nonhuman victims for our prior wrongdoings. Put this way, the hedge might need to stay or go, and the same can be said for the adjacent parking lots, the boat harbor, and even the constructed concrete shoreline itself.

It would be implausible to identify the historic preservationists or birdwatchers as themselves victims of ecological degradation whose preferences should guide ecological reparative justice, though like other community members they may suffer second-order harms in the aftermath. If their voices deserve some special priority, it is not for their own preferences but their informed perspectives on what migratory and resident birds and other negatively affected members of the biotic community need from the city of Chicago (and its leaders, residents, etc.) given our long history of degrading and destroying the shoreline in and around Montrose Point.

The invasive buckthorn that Mary Lou Quinn and others value as a buffer was brought by European migrants to North America to fulfill that human social function, and like honeysuckle it has overrun ecosystems throughout the Chicago Wilderness. Buckthorn and honeysuckle are significant causes of ecosystem degradation, but the perpetrators of environmental harm against Chicago Wilderness ecosystems and their constitutive members are not these invasive plants, of course, but the people who have abetted them. Similarly, the Montrose Point birders, Packard, and Quinn are advocates for migratory birds, tallgrass savanna, and existing trees, respectively, rather than themselves victims of environmental wrongdoing whose needs and priorities should be prioritized in repair. Understanding environmental management in terms of relational repair does not mean blaming invasive species, but instead acknowledging our own culpabilities and working to make amends toward renewed biodiversity and ecosystem recovery.

Second-Order Harms and the Fragility of Trust

Though relatively short-lived, the DuPage and Cook County moratoriums were not without negative ecological consequences, as a report by the Openlands Project emphasized (O'Keefe et al. 2006, 43). While restoration ceased for two to ten years, the rest of the world did not. Other human activities with negative ecological impacts continued, including internal forest preserve projects like trail maintenance and facility construction and external factors like suburban sprawl and coal-fired power plants. Seeding and planting continued at various sites, but without regular prairie burns and invasive-species removal these practices had limited positive effect. Meanwhile invasive species were allowed to retake partially restored sites, undercutting ecosystem recovery and renewed diversity of flowers, grasses, trees, and other plant and animal life there.

For her part restoration critic Carol Nelson told Paddy Woodworth in 2010 that she felt the moratorium had been a mistake:

> "It went on for too long. We should have put a shorter time limit on it, and then invited everyone back to the sites to do surveys, using them as the basis for future management. . . . Jane and John Balaban will never forgive me for supporting it," she says. "I know it sent twenty years of their work down the tubes." (Woodworth 2013, 126)

I had said previously that Packard, Quinn, and their respective allies were not themselves victims whose needs and priorities should guide ecological restoration as relational repair. From animal ethics to epistemic justice, however, a recurring lesson for reparative environmental justice has been that the aftermath of environmental wrongdoing is itself susceptible to further wrongdoing. Sometimes this is because contexts of amelioration are not exempt from the injustices and other wrongs that pervade life generally; sometimes it is because grappling with the consequences of preceding environmental damage and destruction presents its own moral challenges. Here in the Chicago controversy, the fraught relationship between restorationists and their critics included iterative dynamics of second-order harm, distrust, and relational damage.

Consider the issue of trust and its erosion. Packard and his fellow North Branchers earned the trust of FPCC officials through their demonstrated commitment to the work and the approval of recognized experts like Bob Betz. Officials were themselves charged with keeping public trust. The irony is that in their efforts not to draw public criticism, restorationists made certain choices that critics read as duplicitous and thus cause for public concern. "What has bothered me always about restoration in Chicago is the lying," Blix told Woodworth (2013, 126–27). "'If something is that good, [do] you have to lie about it? Trust is completely contaminated in this situation.' And, for her at least, this contamination extends to Chicago Wilderness."[33] By contrast, Lake County Forest Preserve (LCFP) officials suggest that one reason that the restoration moratorium in Cook and DuPage counties did not extend into Lake County, despite opposition to restoration there, was that LCFP was consistently transparent about their activities. Criticism remained, but without cause for distrust.[34]

Once presumptive background trust is called into question, this undercuts restoration aims and practices predicated on that trust.[35] As Coffey (1998g) argued, "If the purported experts running through the Forest Preserves can get it so dead wrong once, how can the taxpayers who pay their salaries be sure they haven't gotten it wrong also on some of those other prairies they're restoring?" For the North Branchers and other dedicated restorationists, meanwhile, Coffey and Quinn did not position themselves as very trustworthy either. We might contrast this antagonistic relationship with interlocutors like Jon Mendelson, who criticized FPCC restoration activities but presented his criticisms as worth engaging. The restorationists were frustrated with Coffey and Quinn not just for their fierce opposition but their seeming inability or unwillingness to seriously engage with the ecology.[36] Quinn saw the forested landscapes of the 1960s as how nature should be, and seemed unable to recognize those forests as themselves owing to resource management. Like Quinn, Coffey emphasized the value of trees given traffic and other pollution sources while dismissing prairie as at best irrelevant to urban air quality. Coffey's skepticism toward controlled burns seemed to be based not only in his distrust of those involved but also a superficial sense of what burns do and how they do it. "Fire kills foreign plant life brought by European immigrants, but selectively spares native American plants?" Coffey (1998a) asked incredulously. "Will only foreign logs burn in American fireplaces? Are these 'restoration' people to be believed?"[37] He actively, openly distrusted restorationists, and they in turn rejected his criticisms as ill-informed and dangerously misleading for a readership that was dependent on his account of the matters in question.[38]

Traditional Ecological Knowledge in the Chicago Wilderness

In their review of *Restoring Nature*, Rodriguez and Fuller defended Chicago restorationists against both hostile opposition and constructive criticism. One such suggestion was from Carol Raish (2000), who urged restorationists to consider how traditional ecological knowledge might play a role in resource management. Rodriguez and Fuller responded as follows:

> Here there are no ancient traditions, no wise ones to emulate, except the volunteers— some practicing for 20 years—and the staff and agencies comprising Chicago Wilderness, less than a decade old. So we are building tradition, adding to our knowledge base, and telling the stories of our successes and failures. (Rodriguez and Fuller 2001, 223)

Rodriguez and Fuller are right to see tradition as something dynamic and ongoing, and yet they cut themselves off at the knees when they characterize Chicago Wilderness as having no ancient traditions and no wise ones to emulate. Chicago is the traditional home of the Ojibwe, Odawa, and Potawatomi Nations, at the confluence of Chicago and Des Plaines rivers and on the shores of Lake Michigan. There are no federally recognized tribes in Illinois today, yet Chicago is home to one of the largest urban American Indian communities in the country, and local organizations such as the American Indian Center and Chi-Nations Youth Council practice native gardening and soil detoxification (American Indian Center 2018; Carrico 2019). Elsewhere in the Chicago Wilderness, the Gun Lake Tribe (2017) and other bands of the Potawatomi Nation in southwest Michigan also conduct native grassland management and environmental restoration projects.

"Traditional ecological knowledge is useful when defining reference ecosystems because Native languages and material culture are a living library of species composition," Kimmerer (2011, 272) argues. "Unfortunately, few ecological scientists are trained to access these valuable data sources." As an ecologist and practicing restorationist herself, Kimmerer is not accusing or scolding. Her argument is that reciprocal restoration offers a way toward becoming *indigenous-to-place*, toward healing our fractured relationships with land and history. "This does not mean appropriating the culture of indigenous people, but generating an authentic new relationship," she explains. "Being indigenous to place means to live as if we'll be here for the long haul, as if our children's future mattered" (ibid.).

In this spirit I would echo the recommendation that giving greater recognition to traditional ecological knowledge and knowers can help resolve the problem in which ecological restoration heals and repairs some relationships while others are further frayed and degraded in the process. This is not a cure-all, of course. It is possible that greater recognition and inclusion of traditional ecological knowledge in the Chicago Wilderness might be reframed by some critics as itself anti-democratic, another form of expert governmentality that is allowed to override local residents' concerns. Yet incorporating traditional knowledge and knowers more fully into both public and private restoration projects serves to strengthen their social-epistemic foundations and expands the communities and relationships that may be repaired in the process.[39]

RESTORE AND REPAIR IN THE CHICAGO WILDERNESS TODAY

I have stressed the importance of centering environmental reparative practices around the needs and perspectives of human and nonhuman victims of environmental injustices and other wrongs. This imperative becomes admittedly difficult to apply, however, when different parties have hurt and undercut each other in the iterative dynamic of social conflicts over restoration.

What is encouraging is that this is a dynamic Chicago Wilderness leaders are aware of. As Heneghan told Woodworth (2013, 87), "I hope we can bring together people who have been mutually suspicious." He identifies his aspirations for the Chicago Wilderness Science Team's Rethinking Ecological and Social Trends of Restoration Ecology (RESTORE) Project this way:

> We want to study this loop—how do people plan, what are the implications for biodiversity, and how do they translate into response by the community? Our thinking is that if planning has gone well, and if the biodiversity outcomes are positive, whatever that might mean, there will be a receptivity in the community for more conservation management. If the planning is poor, if people are excluded from the process, if the plan when put into action has a highly negative impact on the community, then there will not be that receptivity. Our research is rooted in the idea that democratic planning has positive outcomes for both community and environment. (Woodworth 2013, 131)

The Science Team includes natural and social scientists, "designed to produce a more resilient organization . . . better able to interact effectively with managers, policy makers, and the public" (Heneghan et al. 2012, 85). The RESTORE project has so far yielded multiple findings on social-ecological resilience in urban ecological restoration. Here I want to highlight three such findings, which I will call Shared Principles, Similar Outcomes, and Public Attitudes.

Shared Principles

Over the past decade the Science Team has surveyed and interviewed many stakeholders and tracked progress and impediments to ecosystem recovery and biodiversity across multiple sites. From these interviews, Watkins et al. (2015) looked to identify rules, norms, and strategies that guide Chicago Wilderness conservation organizations' decision-making, actions, and outcomes. Recurring themes were aggregated into seven ecological restoration principles: (1) Qualify, don't quantify, (2) Listen to the land, (3) Practice follow-up, (4) Do no harm, (5) Respond to sanctions from the land, (6) Balance diverse internal stakeholders, and (7) Balance diverse external factors (2015, 162–70). The authors note that many of these principles are rooted in deep understanding of the ecological landscape, and furthermore most of them express some sort of norm or strategy. "One advantage of strategies is that they are most readily changed," Watkins et al. (2015, 171) argue, "and therefore are more readily adaptable to new scientific information."[40]

Similar Outcomes

More recently, Heneghan et al. (2019, 2) investigated the following research question: *"Does variation in the organizational structures and decision-making processes that govern the ecological restoration of conservation lands influence outcomes for vegetation structure and diversity?"* They studied ten conservation organizations restoring oak woodlands at fourteen sites in the Chicago region, categorized by three types of social structure: manager-led restorations, co-managed restorations, and researcher-led restorations.[41] Contrary to reasonable expectations, no relationship was found between these categories and vegetation outcomes, which the authors take to be a promising indicator of social resilience in restoration planning.

Public Attitudes

The RESTORE project also included surveys on attitudes and values about restoration among visitors to restored sites and nearby residents, "to investigate the impacts that management styles and resulting biodiversity may have on people not directly involved with ecological restoration, as well as the possible influence of those people on restoration decisions" (Westphal et al. 2014, 20). Gobster et al. (2016) sought to measure public perceptions of urban restoration; beliefs and support were assessed for eight practices common to oak woodland restorations in the Midwest, including native plantings, use of herbicides, controlled burns, and deer control.[42] "As expected, there was much higher support for lower intervention practices like planting and hand weeding, with a general decline in support as the intensity or perceived degree of management intervention increased."[43] There was some variation in support for practices across the surveyed sites, but little difference in support detected between site visitors and neighbors. Interestingly, belief that a restoration practice was being used at a site was the largest predictor of support.[44] "By emphasizing the positive outcomes resulting from management interventions," the authors conclude, "managers might more successfully make the case for their use than if programs such as herbicide spraying or deer removal are communicated without a clear expression of desired outcomes" (Gobster et al. 2016, 223).

Further Lessons

If others of us who live in this damaged and precious biotic community are serious about its social-ecological repair, we must challenge ourselves to contribute to the reparative process in ways that complement and constructively complicate these lessons from the RESTORE project. I suggest that a conceptual framework of reparative environmental justice can help us to do this in several respects. For one, attending to environmental injustice and wrongdoing reminds us that Chicago Wilderness restoration as rectification is at once more specific and wider in scope than just overall improvement in living conditions for human and other members of our community. As valuable as that is, the danger is that we might lose sight of how environmental destruction and degradation have wronged (and continue to wrong) specific groups in our community while others benefited (and continue to benefit) by comparison. This extends beyond present victims, beneficiaries, and perpetrators to include past and future victims, beneficiaries, and perpetrators; beyond those currently living here including those who once lived here, those who have returned, and those who would live here if not for their ancestors' or their own forced migration. How can social-ecological systemic repair take into consideration not only generalized public opinion but specific publics who stand in specific relationship to the past, present, and future of the Chicago Wilderness?

One lesson that we have learned is that distributive justice alone is insufficient for reparative environmental justice; participatory justice and recognition justice matter as well. We must ask how restoration policies and practices benefit some members of our community inequitably, with particular attention to the extent to which such benefits accrue to the victims and perpetrators of prior environmental wrongdoing. Yet prioritizing victim subjectivity also means having a voice in governing restorative practices, and further that one's voice can be heard without smothering or silencing values and knowledge that challenge dominant forms of environmental governance. Public participation in and feedback on restoration projects is consistent with this, but we cannot be content with a passive kind of community engagement, where the general public is welcome to join in and success is measured in terms of total volunteer hours and an absence of critical opposition. If restoration is a way in which we can repair human and nonhuman relationships damaged or destroyed by prior environmental wrongdoing, we must actively seek guidance and renewed interrelationality with those who have been wronged. Restorationists are very much alive to this when it comes to the iterative dynamic of healing one's relationship with the land, and the recognition not just that a site might respond to our reparations but that we must actively invite and listen for such responses (Watkins et al. 2015, 171–72). The RESTORE project shows the need for restorationists and their allies to do the same with neighbors and other members of the general public. Yet we have given too little emphasis in the Chicago Wilderness on the need to go beyond visitors, neighbors, and volunteers, to stronger active partnerships with indigenous groups and other communities of color throughout the region who have been wronged by past and persisting environmental injustices and who remain marginalized in restorationist decision-making and ecological governance more generally.

We have also learned that centering reparative practices on victims' subjectivity is not without challenges. One such challenge arises for contrite perpetrators who want to ground our apologies and amends in victims' needs, values, and perspectives, but then in practice substitute our own extrapolations of what these needs, values, and perspectives might be. We might do this because of our difficulties learning from actual victims, because we are afraid to really listen, or because our hermeneutical ignorance is so far-reaching that we cannot identify or understand them. It is also difficult because genuinely victim-centered relational repair requires vulnerability from all parties, and the histories of injustice calling for reparative justice have undermined the mutual trust needed for even a modest amount of cooperation from perpetrators and victims. As we have seen through close analysis of the buildup and aftermath of the Chicago Wilderness restoration moratorium, this process is further complicated by second-order relational damage and erosions of trust that occur in reparative processes themselves. The challenges involved are considerable, but at least we are not starting from scratch. We can learn from the examples of social-ecological relational repair in our own communities and in other, similarly situated communities. Wronged and marginalized community members have been courageous enough to extend what Horsburgh (1960, 350) calls *therapeutic trust*, voicing their needs and perspectives and making themselves vulnerable, such that in the process perpetrators and beneficiaries of prior injustices might listen, act accordingly, and in doing so show ourselves to be worthy of that trust.

Last but not least, taking a relational approach to reparative environmental justice in the Chicago Wilderness means acknowledging and addressing what Karen Emmerman (2019, 81) calls *moral remainders*, even and especially when such remainders are seemingly unavoidable. We are asked to appreciate the moral weight of culling deer and cutting trees, for example, as morally imperfect responses to a nonideal social-ecological system. *If* deer control is indeed necessary, this is so *given that* we have eliminated (and resist reintroduction of) their natural predators across the region. Alongside moral remainders, we should add Kimmerer's (2013, 15) *allegiance to gratitude*, in which we recognize and demonstrate our appreciation for all we have taken from our biotic communities. We acknowledge the suffering and loss of life to these ends; we do not gloss over or rush past it. We give thanks, and we commit ourselves to live in ways that are worth those sacrifices.

NOTES

1. See Jordan (1986, 2003, 2006), Hourdequin and Wong (2005), Kimmerer (2011, 2013), and Hourdequin (2016).

2. Light (1995, 201); thanks to Jaime Waters for raising the point about wilderness as wasteland.

3. For a critical response to Berry, see Windhager (1998).

4. Quoted in Vena (2013), from Jensen's 1904 report for the Chicago Special Parks Commission.

5. Vena (2013) argues that the FPCC case "exemplifies how legal preservation may save some landscapes, while degrading others." See also Vena (2016) and Heneghan et al. (2012, 78).

6. On Native American uses of fire see Curtis (1959), Dorney and Dorney (1989), and Pyne (2017).

7. Stevens (1995, 38) continues: "They did it partly to provide open spaces where people could stroll, picnic, and relax, partly to keep the brush from spreading, and partly to prevent what some people thought of as an eyesore. Natural savannas and prairies, for all their richness, can seem ragged and disorderly to people who are used to suburban lawns."

8. "To the uninitiated, the idea of a walk through a prairie might seem to be no more exciting than crossing a field of wheat, a cow pasture, or an unmowed blue-grass lawn. Nothing could be further from the truth" (Betz 1972). Packard (1994, 34) said, "The most striking component of Betz's strategy was his growing confidence in the

innate power of the prairie to find its own solutions. Given fire, given time, given a seed source, the prairie would win. Our job is just to help out."

9. North Branch volunteer Preston Sprinks told Stevens (1994, 75), "These guys are wonderful, but they don't know a damned thing about plants."

10. For a skeptical view of the ecological significance of fire, see Mendelson (1998).

11. See also Curtis (1955, 1959) and Packard (1985, 1988).

12. See also Packard (1985, 1993), Stevens (1994, 84–101), and Woodworth (2013, 102).

13. See also Packard (1985) and Woodworth (2013, 101).

14. See also Hyerczyk (1993), Mendelson et al. (1993), Mendelson (1998), Helford (1999), and FPCC (2019).

15. Packard (1993, 10).

16. On habitat restoration, see Yunger (1992) and Palmer et al. (1997); on concerns about habitat loss, see Coffey (1998e) and Woodworth (2013, 125); on animal deaths in prescribed burns, see Stevens (1994, 7) and Strohmaier (2000, 8).

17. Thomas (1991), Stevens (1994, 271–78), Shore (1997, 25), and Woodworth (2013, 113–14).

18. "In an article in a Winnetka newspaper in March 1993, [Gehrie] put forth a concept and a word to describe what must be done: 'We do not want to kill deer, but there they are, crowded into the last wild lands because we have torn apart all the other places. If we are to begin to fix things, some deer must be . . . sacrificed. *Sacrificed*—the giving up of some valued thing for the sake of something of greater value or having a more pressing claim'" (Stevens 1994, 278).

19. See Coffey (1996g, 1996i, 1998a, 1998d) and Rebik (2020).

20. See also Gobster (1997) and Miller (2002).

21. See also Brackett (2013) and Woodworth (2013, 123–31).

22. See Coffey (1996b, 1996c, 1996d, 1996h, 1996j, 1996k, 1998a, 1998c, 1998g, 1998i); see also Shore (1997, 25–26).

23. Coffey (1996b) wrote that "the book made it clear that Packard, glorified as patron saint and principal evangelist of the movement, and his believers realized that not everybody shared their neo-ecological faith." See also Coffey (1998c, 1998i).

24. Packard also said, "We sort of had the authority to do what we could get away with. There was this sort of unwritten agreement: We would do things and get approval in retrospect. If God didn't send lightning bolts down, and the scientists didn't complain, and the public didn't complain, it would be all right" (Stevens 1994, 80).

25. "I asked about the fires and they said, 'Don't worry.' I said, 'But there's houses around.' They said, 'We know what we're doing.' Then I read *Miracle under the Oaks*," Mary Lou Quinn said. "Since then my life has never been the same" (Joravsky 1996).

26. "Packard says that he regards the historic ecosystem not as a literal model for the restorationist but as a 'metaphor' for a healthy, biotically rich, perhaps 'natural' ecosystem" (Jordan and Lubick 2011, 203). See also Packard (2019).

27. Chicago Wilderness (1999) aims "to protect the natural communities of the Chicago region and to restore them to long-term viability, in order to enrich the quality of life of its citizens and to contribute to the preservation of global biodiversity."

Heneghan et al. (2012, 80) argue that the Biodiversity Recovery Plan "was a relatively early adopter of 'ecosystem services' as a valuable framework in which to promote large-scale conservation efforts."

28. See also SER (2020): "The goal of ecological restoration is to return a degraded ecosystem to its historic trajectory, not its historic condition. The ecosystem may not necessarily recover to its former state since contemporary ecological realities, including global climate change, may cause it to develop along an altered trajectory, just as these same realities may have changed the trajectory of nearby undisturbed ecosystems. History plays an important role in restoration, but contemporary conditions must also be taken into consideration."

29. Harris et al. (2006), Seavy et al. (2009), Aronson and Alexander (2013), and Chicago Wilderness (2020).

30. On the relationship between scientific research and restoration practices, see also Chicago Wilderness (1999, 3), Cabin (2007), Heneghan et al. (2009, 63), Gross (2010, 21–22), and Woodworth (2013, 130–31).

31. See also Gobster and Barro (2001).

32. See also Gobster and Barro (2001), Gross (2010, 91), and Lake-Cook Audubon Chapter (2020).

33. Woodworth reports that "I heard from a number of individuals not associated in any way with support for the moratorium that Chicago Wilderness was indeed a cold house for people outside the volunteer movement, or simply with independent spirits, in the early days. . . . No doubt this closed consensus was heavily reinforced by the defensive armour donned by many Chicago Wilderness founders in response to the shock waves created by the moratorium, in a spiral of escalating hurt familiar from so many adversarial political conflicts" (2013, 127).

34. Personal communication with LCFP Director of Natural Resources Jim Anderson and LCFP Chief Operations Officer Mike Tully, February 2020.

35. As birder George Koval put it, "Trust begins to erode. There is very little public input, and I don't think that is the way to win popular support" (in Coffey 1996g).

36. See Friederici (2006, 119), Woodworth (2013, 121–22), and Packard (2016).

37. See also Coffey (1998b, 1998f, 1998i, 1998j, 1998k, 1999).

38. Packard (2016) says that when Debra Shore reached out to Coffey "and tried to engage him with the facts as she'd seen them, along with ideas of journalistic integrity, he shocked and dismissed her with the comment that he didn't need fact-checking, because he was writing columns, and they were just opinions."

39. It is notable that Chicago Wilderness has more than 200 member organizations, but no tribes or other indigenous groups (Chicago Wilderness 2019). For comparison, see Holtgren et al. (2014, 2015).

40. "As the science and practice of ecological restoration advances (setting the stage for creating more rules to guide restoration activities), it is likely that strategies will remain an important institutional type, because of the need to be adaptive and flexible in a dynamic, and at times uncertain, social-ecological system" (Watkins et al. 2015, 171).

41. "At [Manager-led] sites, professional (paid) managers dominated decision-making. On 'Co-managed' sites volunteers worked with professional managers but held a high degree of autonomy in restoration decision-making and management, often leading the decision-making. On 'Researcher-led' sites scientific exploration was central to restoration activities" (Heneghan et al. 2019, 6).

42. Gobster et al. (2016, 219) explicitly build upon Bright et al. (2002) and Miller et al. (2002).

43. Gobster et al. (2016, 221). Support was highest (75–95 percent) for planting, controlled burns, and hand and chainsaw removal; moderate (~50 percent) for removing mature trees and fencing deer; and lowest (30–35 percent) for herbicides and shooting deer.

44. Gobster et al. (2016, 223): "In coding them for the modeling efforts we discovered that it didn't matter whether or not a practice was actually being used at the site—a respondent's belief that they were being used was much more effective in determining their support for it."

Chapter 8

Alone and Together in a World of Wounds

ON THE ROUND RIVER

"One of the penalties of an ecological education," Aldo Leopold writes in *The Round River*, "is that one lives alone in a world of wounds" (1966, 197). In writing this book, I have found myself grappling with every part of this sentence. For one thing, it runs contrary to the value-free ideal central to the positivist conception of science that dominated Leopold's postwar context, that good science sticks to the facts and avoids value judgments.[1] "Wound" is not a value-free observation but rather what Putnam (2002) calls a thick ethical concept, like cruelty, with both descriptive and prescriptive functions. And it is not just a naïve tenderfoot who reads tragedy into the landscape, Leopold insists, in fact ecological education enables (or curses?) one to see the wounds in the world that so many others cannot or will not recognize around them.

But it is more than what we can see—that's too passive, too removed. We *live in* a world of wounds. We drink the water, we breathe the air, we eat cultivated crops and the other creatures to whom we feed the crops, and of course, we spray the crops, burn the coal, and pollute the rivers. We do not just observe the world (though we do that) or change it (though we do that too), but live in it. Living in a world of wounds means recognizing existing wounds and anticipating how our policies and practices will create new ones, and doing all of this not from a position safely removed but always from somewhere within. If this seems too grim, think about "wound" again, built into which is not only damage but the possibility of repair: at least some wounds can be healed. Leopold knew this as well as anyone, from his days as a land manager in the Southwest to his contributions to the field of restoration ecology at the University of Wisconsin Arboretum. It's not that healing is easy or necessarily quick, or that we won't make mistakes, or that we'll be able to dress all the wounds around us. But it gives us direction and serves as a much needed call to action.

And then there is for me the most difficult part of Leopold's observation, that feeling of being alone in this world of wounds. Intimidated and humbled by the reparative task, discouraged and sometimes even despairing. But when Leopold says *alone*, where does leave other ecologists? Where does it leave the rest of the biotic community?

I am reminded here that being alone does not always mean being by oneself, or else familiar sentences like "We are alone in the universe" or "Please, don't leave me alone with him" would not make sense. Perhaps then those with an ecological education are alone together, not so much that each lives in isolation (though that may sometimes be so) but more of a profound distance between them and mainstream society operating according to substantially different assumptions about the world. This kind of lonesome disconnection can be felt throughout *The Round River*. "We who are the heirs of Paul Bunyan have not found out either what we are doing to the river or what the river is doing to us," Leopold (1966, 196) says. "We burl our logs of state with more energy than skill." He is acutely aware of how unaware we so often are, and while there is a kind of ecological Socratic wisdom to this, it can be alienating when those piloting our flotilla down the round river do so with heedless confidence.

"Repairing enduring injustice is not only hard to do," argues Spinner-Halev (2012, 206), "but will often make many uncomfortable." For those who do the work of environmental relational repair—deliberate, reflective practices of animal care, ecological restoration, climate reparations, participatory justice, recognition justice, and so on—does this alleviate or exacerbate Leopold's sense of living alone in a world of wounds? I have argued that reparative environmental justice can help us to break down the moral impediments to healthy relationality created and maintained in the aftermath of environmental wrongdoing. To the extent that our efforts at environmental relational repair succeed, the soil is tilled for repentant perpetrators, forgiving victims, and our communities to reconcile, to live and grow and thrive together, like corn, squash, and beans planted as three sisters.[2]

Akin to Leopold's lonely observation, those doing the work of reparative environmental justice might be discouraged if and when this work seems to happen in a social vacuum, with so many perpetrators of environmental degradation failing to even acknowledge our wrongdoing let alone make international, intergenerational, interspecies, or other amends. Yet even then, if those who do the work of environmental repair are alone, we are alone together. This is, I think, a real strength of a relational approach to environmental justice. Working toward reparative justice is how we close the distance between us that environmental injustice has created or exacerbated.

Reconciliation is far from guaranteed, but environmental reparative justice chips away at the barriers keeping us apart.

WITH OUR FELLOW CREATURES

"So there's another possible solution to the problem of the current great extinction event. We human beings could decide to go extinct," Christine Korsgaard writes in her 2018 book *Fellow Creatures*:

> When I first started looking at the literature on animal ethics, I was a little surprised to find that the friends of animals were not more inclined to advocate this. After all, both the animal ethics literature and the environmental ethics literature endlessly detail the ways in which human beings are the cause of enormous suffering to the other animals, systematically violate their rights, upset the balance of nature and destroy the climate, and so on. We have crowded out the wild animals and filled the world with domestic animals, most of whom are being raised for food and whose short miserable lives are not worth living. It seems especially puzzling that utilitarian authors, who want to maximize happiness and who regard animals as our equals, do not suggest that the members of the species who are knowingly causing all of this misery might have a duty to bow out. But it's the opposite. Singer at one point suggested that on a total utilitarian view, it would be good for the planet to contain as many human beings as it can hold. It is also puzzling that environmental "holists" do not advocate human extinction. Aldo Leopold tells us that "A thing is right when it tends to preserve the integrity, stability, and beauty of the biotic community. It is wrong when it tends otherwise." But nothing has ever been as bad for the biotic community as unhindered human reproduction. Shouldn't it follow that it is wrong for humans to reproduce, and right for us to stop reproducing and let ourselves go extinct? (Korsgaard 2018, 212–13)[3]

Keep in mind that Korsgaard is not asking (why no one is asking) about the question of whether the world *would have been* better off without humans. She is concerned with what we owe our fellow creatures in this world, rather than what we would owe them in one we created from scratch. Given our track record in this world, does morality dictate that things would be better without us?

There are I think at least two not mutually exclusive responses that this literally misanthropic challenge invites. The first is that, as Esme Murdock (2019, 311) says, "it assumes that all human existence is necessarily detrimental to the ecological flourishing of Earth." On the Anthropocene and its connotations, Jeremy Bendik-Keymer (2019) argues that humanity is too broad a referent for who or what is responsible for the aforementioned track record of ecological atrocity.[4] Not all human existence and human practices are rightly impugned. Behind the misanthropic challenge is a settler-colonialist epistemology that marginalizes and actively ignores traditional ecological knowledge, as though all humans have only ever known how to live in dysfunctional antagonism toward other animals and the biotic community.

The second response to the misanthropic challenge foregrounds reparative justice and moral repair. The track record to which Korsgaard draws our attention is itself a reason against ending humanity, rather than a reason for it. Even if the world would have been better had humans never existed, in this world we have and do. Bringing our existence to an end now would at best curtail prospects for further anthropogenic environmental destruction. Yet even that cold comfort is too optimistic, as our nuclear waste, carbon emissions, and invasive-species introductions will make themselves felt long after our collective voluntary end. Once we are gone we cannot do anything to rectify the ecological-ethical implications of our past and present wrongdoings. If we take rectification to mean restoring worldwide ecological conditions to prehuman configurations or compensating the rest of the biotic community for our harmful acts, such grand ambitions will seem hopelessly out of reach. The reparative question is if not easier at least a bit more practical. What do our fellow creatures need from us now, we ask, given how we have wronged them?

LINKED BY SWEETGRASS

Among those impressed by the urgency and enormity of the environmental problems that we have considered—environmental injustice, ecosystem degradation, animal exploitation, global climate change, and epistemic injustice against traditional ecological knowledge—some might worry that reparative environmental justice makes an already imposing task that much more difficult. When the world is on fire, do we really have time to work on relational repair?

Whatever else I have said in this book, I do not mean to suggest that environmental relational repair is easy. The earlier chapters have identified multiple challenges to the contrary, including ways in which reparative processes in the aftermath of environmental wrongdoing themselves can have both positive and negative second-order effects. We also found that prioritizing victim subjectivity provides clarification and guidance for relational repair, yet the overlapping nature of environmental injustices and other wrongs means that perpetrators will owe apologies and amends to multiple victims whose different subjectivities point to different acts of amends and conditions for renewed trust and forgiveness. Meanwhile different perpetrators of collective environmental wrongdoings will also vary in how they participate in reparative processes: able to acknowledge our responsibility for harm or injustices to different degrees, for example, and also more and less willing to make amends for our harmful actions. It can be especially difficult to make sense of the requirements and conditions for renewed trust and forgiveness in interspecies and intergenerational contexts, whether due to difficulties in accessing victims' subjectivities or determining when forgiveness is even an appropriate goal for the relationship in question. All of these theoretical and practical challenges mean that environmental relational repair is not an all-or-nothing affair. Some relationships are more easily repaired than others, and even then, environmental reparative justice is not a threshold reached but an ongoing process of becoming morally healthier, more trustworthy, and resilient.

The urge to find environmental solutions unhampered by the work of relational repair steers us toward shortcuts filled with pitfalls, impediments, and detours of their own. As Whyte (2020, 3) argues on global climate change and indigenous environmental justice, prioritizing ecological problems without regard for relational ones is bound to perpetuate and exacerbate unresolved injustices. It also means otherwise promising routes to addressing serious ecological problems will be closed off to us without also addressing our unresolved relational problems: solutions requiring interspecies, intercultural, international, and intergenerational collaborations that are undermined by our inadequately addressed histories of environmental wrongdoing. This will be true for perpetrators who want to move forward without doing the difficult, humbling work of reparative justice, since without doing this work they cannot demonstrate ethical and epistemic growth and renewed trustworthiness, which enable victims and active bystanders to collaborate with them. It will also be true for victims, though for different reasons. The problem is not that they must demonstrate trustworthiness to those who have wronged them; indeed, some projects of ecological, interspecies, and community repair may be possible without having to collaborate with untrusted offenders. But given their power and influence, many solutions to environmental problems remain out of reach so long as perpetrators are (knowingly or unknowingly) aligned against them.

"A sheaf of sweetgrass, bound at the end and divided into thirds, is ready to braid," writes Robin Kimmerer. She continues:

> Of course you can do it yourself . . . but the sweetest way is to have someone else hold the end so that you pull gently against each other, all the while leaning in, head to head, chatting and laughing, watching each other's hands, one holding steady while the other shifts the slim bundles over one another, each in its

turn. Linked by sweetgrass, there is reciprocity between you, linked by sweetgrass, the holder as vital as the braider. The braid becomes finer and thinner as you near the end, until you're braiding individual blades of grass, and then you tie it off. (Kimmerer 2013, ix)

One cannot weave a strong braid with only two strands. The third is not in itself more important, but nothing lasting can be braided without it. The flowers cannot cross-pollinate without the bees dancing between them, any more than the bees can do the work alone in the sky with nothing to gather and nowhere to land.

The time and fortitude needed to weave us together in the aftermath of environmental wrongs are not insignificant. A rush job will not do, no matter how urgent our situation. Quickly twisting unrepentant perpetrators and unacknowledged victims together might look like a decent fix, but we will unravel soon enough. We need that third strand, which on this metaphor can be the larger community or perhaps the reparative process itself. Even then, we must recognize the possibility that one or more of our strands may not take a braid. Environmental relational repair is not easy, and sometimes not even possible: perhaps because the reparative process is rushed, perpetrators cannot or will not do the work, victims cannot or will not open themselves to more pain, or some combination of these. But when conditions allow for reparative environmental justice, and we do the work of environmental relational repair with critical reflection, sensitivity, and reciprocity, we are capable of more braided together than left frayed and alone.

NOTES

1. On the value-free ideal for science, see Kincaid et al. (2007), Douglas (2009), and Brown (2020).

2. "Being among the sisters provides a visible manifestation of what a community can become when its members under-stand and share their gifts," Kimmerer (2013, 134) writes. "In reciprocity, we fill our spirits as well as our bellies."

3. We might think of Korsgaard's provocation as an environmental spin on Camus's existential claim that the one truly serious philosophical problem is that of suicide. The philosophy blog *Existential Comics* offers a similar eco-misanthropist analysis through the voice of Lovecraft's terrifying creation Cthulhu: "Come on! You are currently destroying the planet, and for what? So you can have more TVs? More cars? More beef? More money? It's obvious that humanity causes more suffering than happiness." His human interlocutor concedes that Cthulhu makes a pretty good point: http://existentialcomics.com/comic/314.

4. See also Sharp (2020).

Bibliography

"A Place at the Table." 1993. *Sierra Magazine*, May/June: 51–8, 90–1.

Aaltola, Elisa. 2005. "Animal Ethics and Interest Conflicts." *Ethics and the Environment* 10(1): 19–48.

Abbate, Cheryl. 2016. "How to Help When It Hurts." *Journal of Social Philosophy* 47: 142–70.

Acorn, Annalise. 2012. "The Seductive Vision of Restorative Justice." In *A Restorative Justice Reader*, edited by Gerry Johnstone. New York: Routledge.

Adams, Carole J. 1990. *The Sexual Politics of Meat*. New York: Continuum.

Adams, Carole J. 1994. *Neither Man nor Beast*. New York: Continuum.

Agrawal, Arun. 1995. "Dismantling the Divide Between Indigenous and Scientific Knowledge." *Development and Change* 26(3): 413–39.

Agyeman, Julian, David Schlosberg, Luke Craven, and Caitlin Mathews. 2016. "Trends and Directions in Environmental Justice." *Annual Review of Environment & Resources* 41: 321–40.

Agyeman, Julian, Peter Cole, Randolph Haluza-DeLay, and Pat O'Riley, eds. 2009. *Speaking for Ourselves: Environmental Justice in Canada*. Vancouver: UBC Press.

Alario, Margarita. 2000. "Urban and Ecological Planning in Chicago." *Journal of Environmental Planning and Management* 43(4): 489–504.

Allison, Stuart. 2004. "What Do We Mean When We Talk About Restoration?" *Ecological Restoration* 22(4): 281–86.

Allison, Stuart. 2007. "You Can't Not Choose." *Restoration Ecology* 15(4): 601–5.

Almassi, Ben. 2012a. "Climate Change, Epistemic Trust, and Expert Trustworthiness." *Ethics and the Environment* 17(2): 29–49.

Almassi, Ben. 2012b. "Climate Change and the Ethics of Individual Emissions." *Perspectives* 4: 4–21.

Almassi, Ben. 2016. "Toxic Funding? Conflicts of Interest and Their Epistemological Significance." *Journal of Applied Philosophy* 34(2): 206–20.

Almassi, Ben. 2018. "What's Wrong with Ponzi Schemes?" *International Journal of Applied Philosophy* 32(1): 111–26.

American Indian Center. 2020. "History." https://aicchicago.org/history/.
Anderson, Allen, and Linda Anderson. 2006. *Rescued.* Novato, CA: New World Library.
Anderson, Elizabeth. 2010. *The Imperative of Integration.* Princeton: Princeton University Press.
Anderson, Elizabeth. 2012. "Epistemic Justice as a Virtue of Social Institutions." *Social Epistemology* 26: 163–73.
Aristotle. 1990. *Nicomachean Ethics*, translated by David Ross. Oxford: Oxford University Press.
Aronson, James, and Sasha Alexander. 2013. "Ecological Restoration Is Now a Global Priority." *Restoration Ecology* 21: 293–6.
Baatz, Christian. 2013. "Responsibility for the Past?" *Ethics, Policy, and Environment* 16: 94–110.
Baier, Annette. 1981. "The Rights of Past and Future Persons." In *Responsibilities to Future Generations*, edited by Ernest Partridge, 171–83. New York: Prometheus Books.
Baier, Annette. 1986. "Trust and Anti-Trust." *Ethics* 96(2): 231–60.
Bannister, Kelly, Maui Solomon, and Conrad Brunk. 2009. "Appropriation of Traditional Knowledge." In *The Ethics of Cultural Appropriation*, edited by James Young and Conrad Brunk. Oxford: Wiley-Blackwell.
Barnett, Randy. 2013. "Restitution: A New Paradigm of Criminal Justice." In *A Restorative Justice Reader*, edited by Gerry Johnstone. New York: Routledge.
Barrett, Steven, Raymond Speth, Sebastian Eastham, Irene Dedoussi, Akshay Ashok, Robert Malina, and David Keith. 2015. "Impact of the Volkswagen Emissions Control Defeat Device on US Public Health." *Environmental Research Letters* 10(11): 114005.
Barro, Susan, and Alan Bright. 1998. "Public Views on Ecological Restoration." *Restoration & Management Notes* 16(2): 59–65.
Barry, Brian. 1999. "Sustainability and Intergenerational Justice." In *Fairness and Futurity*, edited by Andrew Dobson, 93–117. Oxford: Oxford University Press.
Basl, John. 2010. "Restitutive Restoration." *Environmental Ethics* 32(2): 135–47.
Bass, Gary. 2012. "Reparations as a Noble Lie." *Nomos* 51: 166–79.
Battiste, Marie Ann, and James Youngblood Henderson. 2000. *Protecting Indigenous Knowledge and Heritage.* Vancouver: UBC Press.
Been, Virginia. 1992. "What's Fairness Got to Do with It?" *Cornell Law Review* 78: 1001.
Been, Virginia. 1995. "Analyzing Evidence of Environmental Justice." *Journal of Land Use and Environmental Law* 11(1): 1–28.
Bendik-Keymer, Jeremy. 2019. "Autonomous Conceptions of our Planetary Situation." *International Society for Environmental Ethics Summer Conference*, Blue River, OR.
Berenstain, Nora. 2016. "Epistemic Exploitation." *Ergo* 33(2): 569–90.
Berkes, Fikret. 1999. *Sacred Ecology.* Philadelphia: Taylor & Francis.
Berlinger, Nancy. 2005. *After Harm.* Baltimore: Johns Hopkins University Press.
Bernton, Hal, and Christine Clarridge. 2006. "Earth Liberation Front Members Plead Guilty in 2001 Fire-Bombing." *Seattle Times*, September 24.

Berry, Dwight. 1998. "Toward Reconciling the Cultures of Wilderness and Restoration." *Restoration & Management Notes* 16(2): 125–7.
Betz, Robert. 1972. "What is a Prairie?" In *The Prairie: Swale and Swell,* edited by Torkel Korling. Dundee, IL. Available online: http://www.plantconservation.us/Betz1972.pdf.
Betz, Robert. 1986. "One Decade of Research in Prairie Restoration at the Fermi National Accelerator Laboratory Batavia IL." *Proceedings of the North American Prairie Conference.*
Betz, Robert, Robert Lootens, and Michael Becker. 1996. "Two Decades of Prairie Restoration at Fermilab, Batavia IL." *Proceedings of the North American Prairie Conference.*
Biggar, Nigel. 2001. "Making Peace or Doing Justice: Must We Choose?" In *Burying the Past,* edited by Nigel Biggar. Washington, DC: Georgetown University Press.
Blix, Petra. 2007. "Deafening Silence Over the Assault on 'Brush'." *Chicago Tribune,* January 15.
Blustein, Jeffrey. 2015. "How the Past Matters." In *Historical Justice and Memory,* edited by Klaus Newman and Janna Thompson. Madison: University of Wisconsin Press.
Bondy, Patrick. 2010. "Argumentative Injustice." *Informal Logic* 30(3): 263–78.
Bostock, Stephen St. C. 1993. *Zoos and Animal Rights.* New York: Routledge.
Boxill, Bernard. 1972. "The Morality of Reparations." *Social Theory and Practice* 2: 113–23.
Brackett, Elizabeth. 2013. "Forest Preserve Restoration." *WTTW,* November 6.
Braithwaite, John. 1999. "Restorative Justice." *Crime and Justice* 25: 1–127.
Bright, Alan, Susan Barro, and Randall Burtz. 2002. "Public Attitudes Toward Ecological Restoration in the Chicago Metropolitan Region." *Society and Natural Resources* 15: 763–85.
Broome, John. 2019. "Against Denialism." *The Monist* 102(1): 110–29.
Brown, Jennifer. 2012. *The Lucky Ones.* New York: Avery.
Brown, Matthew. 2020. *Science and Moral Imagination.* Pittsburgh: University of Pittsburgh Press.
Brugge, Doug, Jamie deLemos, and Cat Bui. 2007. "The Sequoyah Corporation Fuels Release and the Church Rock Spill." *American Journal of Public Health* 97: 1595–600.
Brugge, Doug, Timothy Benally, and Ester Yazzie-Lewis. 2006. *The Navajo People and Uranium Mining.* Albuquerque: University of New Mexico Press.
Brulle, Robert, and David Pellow. 2006. "Environmental Justice: Human Health and Environmental Inequalities." *Annual Review of Public Health* 27: 103–24.
Buckley, Madeline. 2020. "In Illinois – The Prairie State – Little Prairie Land Remains." *Chicago Tribune,* February 6.
Bullard, Robert D., ed. 1993. *Confronting Environmental Racism.* Boston: South End Press.
Bullard, Robert D. 1994. *Dumping in Dixie.* Boulder: Westview Press.
Bullard, Robert D. 2001. "Environmental Justice in the 21st Century." *Phylon* 49(3–4): 151–71.

Bullard, Robert D., Paul Mohai, Robin Saha, and Beverly Wright. 2007. *Toxic Wastes and Race at Twenty: 1987–2007*. United Church of Christ, Commission on Racial Justice.

Burgess Jackson, Keith. 1998. "Doing Right by Our Animal Companions." *Journal of Ethics* 2: 159–85.

Butt, Daniel. 2015. "Historical Justice in Political Contexts." In *Historical Justice and Memory*, edited by Klaus Newman and Janna Thompson. Madison: University of Wisconsin Press.

Cabin, Robert. 2007. "Science-Driven Restoration." *Restoration Ecology* 15: 1–7.

Cahn, Robert, ed. 1985. *An Environmental Agenda for the Future*. Washington, DC: Island Press.

Cairns, John. 1995. "Eco-Societal Restoration: Reestablishing Humanity's Relationship with Natural Systems." *Environment: Science and Policy for Sustainable Development* 37(5): 4–33.

Cairns, John. 2003. "Ethical Issues in Ecological Restoration." *Ethics in Science and Environmental Politics* 3: 50–61.

Cairns, John, and John Heckman. 1996. "Restoration Ecology: The State of an Emerging Field." *Annual Review of Energy and Environment* 21: 167–89.

Callicott, J. Baird. 1988. "Animal Liberation and Environmental Ethics: Back Together Again." *Between the Species* 4(3): 163–9.

Caney, Simon. 2012. "Just Emissions." *Philosophy and Public Affairs* 40(4): 255–300.

Caney, Simon. 2016. "Climate Change and Non-Ideal Theory." In *Climate Justice in a Non-Ideal World*, edited by Clare Heyward and Dominic Roser. Oxford: Oxford University Press.

Cannon, Mary. 2007. "If You Build It: Reclaiming Rollins Savanna." *Habitat Herald* 8(1): 1–4.

Capek, Stella. 1993. "The 'Environmental Justice' Frame." *Social Problems* 40(1): 5–24.

Card, Claudia. 2004. "The Atrocity Paradigm Revisited." *Hypatia* 19(4): 210–20.

Carrico, Natalya. 2019. "'We're Still Here': The First Nations Garden in Albany Park Aims to Heal the Community and the Environment." *Chicago Reader*, March 19.

Ceccarelli, Leah. 2011. "Manufactured Scientific Controversy." *Rhetoric and Public Affairs* 14(2): 195–228.

Chapman, Robert. 2006. "Ecological Restoration Restored." *Environmental Values* 15: 463–78.

Chavis, Benjamin, and Charles Lee. 1987. *Toxic Wastes and Race in the United States*. United Church of Christ, Commission on Racial Justice.

Chicago Wilderness. 1999. "Chicago Wilderness and its Biodiversity Recovery Plan." https://cdn.ymaws.com/www.chicagowilderness.org/resource/resmgr/Publications/biodiversity_recovery_plan.pdf.

Chicago Wilderness. 2016. "Native Landscape and Ecological Restoration Guide." https://cdn.ymaws.com/www.chicagowilderness.org/resource/resmgr/publications/Native_Landscaping_Guide.pdf.

Chicago Wilderness. 2019. "Our Partners." www.chicagowilderness.org/page/OurPartners.
Chicago Wilderness. 2020. "Climate Action." www.chicagowilderness.org/page/climate_action.
Choi, Young. 2005. "Theories for Ecological Restoration in a Changing Environment: Toward Futuristic Restoration." *Ecological Research* 19: 75–81.
Chossiere, Guillaume, Robert Malina, Akshay Ashok, Irene Dedoussi, Sebastian Eastham, Raymond Speth, and Steven Barrett. 2017. "Public Health Impacts of Excess NOX Emissions from Volkswagen Diesel Passenger Vehicles in Germany." *Environmental Research Letters* 12(3): 1–14.
Ciokajlo, Mickey. 2006. "Burn Moratorium Lifted at 5 Forest Preserve Sites." *Chicago Tribune*, October 4.
Clewall, Andre. 2000. "Restoring for Natural Authenticity." *Ecological Restoration* 18: 216–17.
Clifford, William K. 1886. *Lectures and Essays*. London: MacMillan.
Coady, David. 2010. "Two Concepts of Epistemic Injustice." *Episteme* 7(2): 101–13.
Coady, David. 2017. "Epistemic Injustice as Distributive Injustice." In *The Routledge Handbook of Epistemic Injustice*, edited by Ian James Kidd, Gaile Pohlhaus Jr, and Jose Medina. London: Routledge.
Coady, David, and Richard Corry. 2013. *The Climate Change Debate*. New York: Palgrave.
Coffey, Ray. 1996a. "Half-Million Trees Face Ax." *Chicago Sun-Times*, May 12.
Coffey, Ray. 1996b. "Forest District 'Partners' Have Shady History." *Chicago Sun-Times*, May 31.
Coffey, Ray. 1996c. "Restorationists Gnaw at Forest Picnic Area." *Chicago Sun-Times*, June 4.
Coffey, Ray. 1996d. "Restorationists Talk a Chic, But Vague, Game." *Chicago Sun-Times*, June 7.
Coffey, Ray. 1996e. "Lake County New Front Line in Forest Preserve War." *Chicago Sun-Times*, June 20.
Coffey, Ray. 1996f. "County Board Gets Wake Up Call on Forest Preserves." *Chicago Sun-Times*, September 24.
Coffey, Ray. 1996g. "Time to Restore Decorum to Forest Preserve Debate." *Chicago Sun-Times*, October 10.
Coffey, Ray. 1996h. "Forest Preserve Brass Shrugs Off Tree Hit List." *Chicago Sun-Times*, October 11.
Coffey, Ray. 1996i. "Forest Preserve Debate Going Before the Public." *Chicago Sun-Times*, October 29.
Coffey, Ray. 1996j. "Forest Preserves' 'Controlled Fires' Raise Concerns." *Chicago Sun-Times*, October 31.
Coffey, Ray. 1996k. "Forest Preserve District is Picking Our Poison." *Chicago Sun-Times*, November 1.
Coffey, Ray. 1997. "Forest Preserves at Center of Controversy–Again." *Chicago Sun-Times*, April 11.

Coffey, Ray. 1998a. "Smoking Out the County's Tree-Burning Plan." *Chicago Sun-Times*, February 20.
Coffey, Ray. 1998b. "Forest District Trees Still a Burning Issue." *Chicago Sun-Times*, March 6.
Coffey, Ray. 1998c. "Guru's Forest Restoration Plans Read More Like Destruction." *Chicago Sun-Times*, March 10.
Coffey, Ray. 1998d. "Young Conscripts Aid Forest Preserve Zealots." *Chicago Sun-Times*, March 24.
Coffey, Ray. 1998e. "Bird Lovers See Threat in Prairie Restoration." *Chicago Sun-Times*, March 26.
Coffey, Ray. 1998f. "Tree-Burning Creates New Ecosystem, Foes Say." *Chicago Sun-Times*, March 27.
Coffey, Ray. 1998g. "An Embarrassing Admission from Prairie 'Experts.'" *Chicago Sun-Times*, April 28.
Coffey, Ray. 1998h. "Trees Were Sacrificed to Print Some Egghead's Tract." *Chicago Sun-Times*, July 7.
Coffey, Ray. 1998i. "'Restoration' Fires Are Wild by Nature." *Chicago Sun-Times*, July 9.
Coffey, Ray. 1998j. "Keep Fires Burning Close to Home." *Chicago Sun-Times*, August 13.
Coffey, Ray. 1998k. "Burning Issues at Nature Center." *Chicago Sun-Times*, November 27.
Coffey, Ray. 1999. "Out of Control Burn Should Teach a Lesson." *Chicago Sun-Times*, November 4.
Cole, Luke, and Sheila Foster. 2001. *From the Ground Up: Environmental Racism and the Rise of the Environmental Justice Movement*. New York: New York University Press.
Collin, Robin, and Robert Collin. 2005. "Environmental Reparations." In *The Quest for Environmental Justice*, edited by Robert D. Bullard. San Francisco: Sierra Club Books.
Congdon, Daniel Avery. 2014. *Restoration as Caring Practice: A Relational Perspective on Ecological Recovery*. Graduate Student Theses, Dissertations, & Professional Papers 4352.
Consedine, Jim. 1995. "The Maori Restorative Tradition." In *A Restorative Justice Reader*, edited by Gerry Johnstone. Cullompton: Willan Publishing.
Cook, Anna. 2016. "A Politics of Indigenous Voice: Reconciliation, Felt Knowledge, and Settler Denial." *The Canadian Journal of Native Studies* 36(2): 69–80.
Cottingham, John. 1979. "Varieties of Retribution." *The Philosophical Quarterly* 29: 238–46.
Crawford, Colin. 1996. "Analyzing Evidence of Environmental Justice." *Journal of Land Use and Environmental Law* 12(1): 103–20.
Crisp, Terri. 2000. *Emergency Animal Rescue Stories*. Roseville, CA: Prima Publishing.
Cronon, William. 1996. "The Trouble with Wilderness, or Getting Back to the Wrong Nature." *Ecological History* 1(1): 7–28.

Curtin, Deane. 1991. "Toward an Ecological Ethic of Care." *Hypatia* 6(1): 60–74.
Curtis, John T. 1955. "A Prairie Continuum in Wisconsin." *Ecology* 96(4): 558–66.
Curtis, John T. 1959. *The Vegetation of Wisconsin*. Madison: University of Wisconsin Press.
Cutter, Susan. 2006. *Hazards, Vulnerability, and Environmental Justice*. New York: Earthscan.
Daly, Kathleen. 2003. "Restorative Justice: The Real Story." In *A Restorative Justice Reader*, edited by Gerry Johnstone. Cullompton: Willan Publishing.
Daly, Kathleen. 2006. "The Limits of Restorative Justice." In *Handbook of Restorative Justice*, edited by Dennis Sullivan and Larry Tifft. London: Routledge.
Davis, Emmalon. 2016. "Typecasts, Tokens, and Spokespersons." *Hypatia* 31(2): 1–17.
Davis, Emmalon. 2018. "On Epistemic Appropriation." *Ethics* 128(4): 702–27.
D'Costa, Russell C. 2005. "Reparations as a Basis for the Makah's Right to Whale." *Animal Law* 12: 71.
De Grieff, Pablo. 2006. "Justice in Reparations." In *The Handbook on Reparations*, edited by Pablo de Grieff, 390–419. Oxford: Oxford University Press.
De Melo-Martin, Inmaculada, and Kristen Intemann. 2018. *The Fight Against Doubt*. Oxford: Oxford University Press.
De Shalit, Avner. 1995. *Why Posterity Matters*. New York: Routledge.
De Shalit, Avner. 2011. "Climate Change Refugees, Compensation, and Rectification." *The Monist* 94(3): 310–28.
Diller, Lowell. 2013. "To Shoot or Not to Shoot." *The Wildlife Professional*, Winter.
Donaldson, Sue, and Will Kymlicka. 2011. *Zoopolis*. Oxford: Oxford University Press.
Donaldson, Sue, and Will Kymlicka. 2015. "Farmed Animal Sanctuaries: The Heart of the Movement?" *Politics and Animals* 1: 50–74.
Dorney, Cheryl H., and John R. Dorney. 1989. "An Unusual Oak Savanna in Northeastern Wisconsin: The Effect of Indian-Caused Fire." *American Midland Naturalist* 122(1): 103–13.
Dotson, Kristie. 2011. "Tracking Epistemic Violence, Tracking Practices of Silencing." *Hypatia* 26: 236–57.
Dotson, Kristie. 2012. "A Cautionary Tale." *Frontiers* 33: 24–47.
Douglas, Heather. 2009. *Science, Policy, and the Value-Free Ideal*. Pittsburgh: University of Pittsburgh Press.
Doyle, Catherine. 2017. "Captive Wildlife Sanctuaries." *Animal Studies Journal* 6(2): 55–85.
Dunayer. 2004. *Speciesism*. Derwood, MD: Ryce Publishing.
Durlin, Marty. 2010. "The Shot Heard Round the West." *High Country News* 42(2), February 1.
Egan, Dave. 2001. "Beyond Members." *Ecological Restoration* 19(2): 67–8.
Egan, Dave. 2006. "Authentic Ecological Restoration." *Ecological Restoration* 24: 223–4.

Egan, Dave, Evan Hjerpe, and Jesse Abrams. 2011. "Why People Matter in Ecological Restoration." In *Human Dimensions of Ecological Restoration*, edited by Dave Egan, Evan Hjerpe, and Jesse Abrams. Washington, DC: Island Press.
El-Hani, Charbel Nino, and Fabio Pedro Souza de Ferreira Bandeira. 2008. "Valuing Indigenous Knowledge." *Cultural Studies of Science Education* 3: 751–79.
Elliot, Robert. 1982. "Faking Nature." *Inquiry* 25(1): 81–93.
Elliot, Robert. 1994. "Extinction, Restoration, Naturalness." *Environmental Ethics* 16: 135–44.
Elliot, Robert. 1997. *Faking Nature*. London: Routledge.
Emerick, Barrett. 2017. "Forgiveness and Reconciliation." In *The Moral Psychology of Forgiveness*, edited by Kathryn J. Norlock, 117–34. London: Rowman & Littlefield.
Emmerman, Karen. 2014a. "Sanctuary, Not Remedy: Inter-Animal Moral Repair and the Problem of Captivity." In *Ethics of Captivity*, edited by Lori Gruen. Oxford: Oxford University Press.
Emmerman, Karen. 2014b. "Inter-Animal Moral Conflicts and Moral Repair." In *Ecofeminism*, edited by Carole Adams and Lori Gruen. New York: Bloomsbury.
Emmerman, Karen. 2019. "What's Love Got to Do with It? An Ecofeminist Approach to Inter-Animal and Inter-Cultural Conflicts of Interest." *Ethical Theory and Moral Practice* 22: 77–91.
Fegelman, Andrew. 1996. "Restoration of Area Prairie Put on Hold." *Chicago Tribune*, September 25.
Figueroa, Robert. 2001. "Latinos and Environmental Justice." In *Faces of Environmental Racism*, edited by Laura Westra and Bill Lawson. Lanham: Rowman & Littlefield.
Figueroa, Robert. 2006. "Evaluating Environmental Justice Claims." In *Forging Environmentalism*, edited by Joanne Bauer. New York: M.E. Sharpe.
Figueroa, Robert, and Claudia Mills. 2001. "Environmental Justice." In *A Companion to Environmental Philosophy*, edited by Dale Jamieson. Oxford: Blackwell Publishers.
Figueroa, Robert, and Gordon Waitt. 2008. "Cracks in the Mirror: (Un)Covering the Moral Terrains of Environmental Justice at Uluru-Kata Tjuta National Park." *Ethics, Place, and Environment* 11(3): 327–49.
Figueroa, Robert, and Gordon Waitt. 2010. "Climb: Restorative Justice, Environmental Heritage, and the Moral Terrains of Uluru-Kata Tjuta National Park." *Journal of Environmental Philosophy* 7(2): 135–63.
First National People of Color Environmental Leadership Summit. 1991. "Principles of Environmental Justice." https://www.ejnet.org/ej/principles.html.
Forest Preserves of Cook County (FPCC). 2019. "Cap Sauers Holding Nature Preserve." https://fpdcc.com/places/locations/cap-sauers-holding-nature-preserve/.
Forest Preserves of Cook County (FPCC). 2020. "Mission & History." fpdcc.com/about/mission-history/.
France, Robert L. 2008. "Introduction." In *Healing Natures, Repairing Relationships*, edited by Robert L. France. Sheffield: Green Frigate Books.
Francione, Gary. 2000. *Introduction to Animal Rights*. Philadelphia: Temple University Press.

Francione, Gary. 2005. *Rain Without Thunder.* Philadelphia: Temple University Press.
Francione, Gary. 2009. *Animals as Persons.* New York: Columbia University Press.
Fredericks, Sarah. 2011. "Monitoring Environmental Justice." *Environmental Justice* 4(1): 63–9.
Fredericks, Sarah. 2019. "Climate Apology and Forgiveness." *Journal of the Society of Christian Ethics* 39(1): 143–59.
Fricker, Miranda. 1998. "Rational Authority and Social Power: Towards a Truly Social Epistemology." *Proceedings of the Aristotelian Society* 98(2): 159–77.
Fricker, Miranda. 2006. "Powerlessness and Social Interpretation." *Episteme* 3(1–2): 96–108.
Fricker, Miranda. 2007. *Epistemic Injustice.* Oxford: Oxford University Press.
Friederici, Peter. 2006. *Nature's Restoration.* Washington, DC: Island Press.
Gaard, Greta. 1997. "Toward a Queer Ecofeminism." *Hypatia* 12(1): 114–37.
Gaard, Greta. 2001. "Tools for a Cross-cultural Feminist Ethics: Exploring Ethical Contexts and Contents in the Makah Whale Hunt." *Hypatia* 16(1): 1–26.
Gardiner, Stephen M. 2009. "A Contract on Future Generations?" In *Intergenerational Justice*, edited by Axel Gosseries and Lukas Meyer, 77–118. Oxford: Oxford University Press.
Gardiner, Stephen M. 2011. *A Perfect Moral Storm: The Ethical Tragedy of Climate Change.* Oxford: Oxford University Press.
Gardiner, Stephen M., and Lauren Hartzell-Nichols. 2012. "Ethics and Global Climate Change." *Nature Education Knowledge* 3(10): 5.
Garvey, James. 2008. *The Ethics of Climate Change.* London: Continuum.
Geider, Ken, and Gerry Waneck. 1983/1994. "PCBs and Warren County." Reprinted in *Unequal Protection*, edited by Robert Bullard. San Francisco: Sierra Club Books.
Gelfert, Axel. 2013. "Climate Skepticism, Epistemic Dissonance, and the Ethics of Uncertainty." *Philosophy and Public Issues* 3(1): 167–208.
Gibson, Julia D. 2019. *Climate Justice for the Dead and Dying.* Dissertation, Michigan State University.
Gilbert, Samuel. 2019. "Church Rock, America's Forgotten Nuclear Disaster, Is Still Poisoning Navajo Lands 40 Years Later." *Vice*, August 12.
Gilio-Whitaker, Dina. 2019. *As Long as Grass Grows: The Indigenous Fight for Environmental Justice from Colonization to Standing Rock.* Boston: Beacon Press.
Gilson, Erinn. 2011. "Vulnerability, Ignorance, and Oppression." *Hypatia* 26(2): 308–32.
Glen, Samantha. 2001. *Best Friends.* New York: Kensington Books.
Gobster, Paul. 1994. "The Urban Savanna." *Restoration & Management Notes* 12(1): 64–71.
Gobster, Paul. 1997. "The Other Side." *Restoration & Management Notes* 15(1): 32–7.
Gobster, Paul. 2000. "Introduction." In *Restoring Nature: Perspectives from the Social Sciences and Humanities*, edited by Paul Gobster and R. Bruce Hall. Washington, DC: Island Press.

Gobster, Paul. 2010. "Introduction: Urban Ecological Restoration." *Nature and Culture* 5: 227–30.

Gobster, Paul, and Susan Barro. 2000. "Negotiating Nature." In *Restoring Nature*, edited by Paul Gobster and R. Bruce Hall. Washington, DC: Island Press.

Gobster, Paul, Kristin Floress, Lynne Westphal, Cristy Watkins, Joanne Vining, and Alaka Wali. 2016. "Resident and User Support for Urban Natural Areas Restoration Practices." *Biological Conservation* 203: 216–25.

Goldman, Benjamin, and Laura Fitton. 1994. *Toxic Wastes and Race Revisited*. Washington, DC: Center for Policy Alternatives.

Goodin, Robert. 1985. *Protecting the Vulnerable*. Chicago: University of Chicago Press.

Grasswick, Heidi. 2017. "Epistemic Injustice in Science." In *Routledge Handbook of Epistemic Injustice*, edited by Ian James Kidd, Gaile Pohlhaus Jr, and Jose Medina. London: Routledge.

Greenburg, Julie. 2014. "Beyond Allyship." *Tikkun* 29(1): 11–7.

Gross, Matthias. 2010. *Ignorance and Surprise: Science, Society, and Ecological Design*. Cambridge: MIT Press.

Gruen, Lori. 1993. "Dismantling Oppression: An Analysis of the Connection Between Women and Animals." In *Ecofeminism*, edited by Greta Gaard. Philadelphia: Temple University Press.

Gruen, Lori. 1996. "On the Oppression of Women and Animals." *Environmental Ethics* 18(4): 441–4.

Gruen, Lori. 2009. "Attending to Nature." *Ethics and the Environment* 14(2): 23–38.

Gruen, Lori. 2015. *Entangled Empathy*. New York: Lantern Books.

Gun Lake Tribe. 2017. "Environmental Projects." https://gunlaketribe-nsn.gov/departments/administration/environmental/environmental-projects.

Gunasekera, Crystal. 2018. "The Ethics of Killing 'Surplus' Zoo Animals." *Journal of Animal Ethics* 8(1): 93–102.

Gunn, Alastair. 1991. "The Restoration of Species and Natural Environments." *Environmental Ethics* 13: 291–310.

Haalboom, Bethany, and David Natcher. 2012. "The Power and Peril of 'Vulnerability'." *Artic* 65: 319–27.

Hampton, Jean. 1984. "Moral Education Theory of Punishment." *Philosophy & Public Affairs* 13: 208–38.

Haraway, Donna. 2008. *When Species Meet*. Minneapolis: University of Minnesota Press.

Hardwig, John. 1991. "The Role of Trust in Knowledge." *Journal of Philosophy* 88: 293–308.

Harris, James, Richard Hobbs, Eric Higgs, and James Aronson. 2006. "Ecological Restoration and Global Climate Change." *Restoration Ecology* 14(2): 170–76.

Heath, Joseph. 2013. "The Structure of Intergenerational Cooperation." *Philosophy & Public Affairs* 41(1): 31–66.

Hedberg, Trevor. 2016. "Animals, Relations, and the Laissez-Faire Intuition." *Environmental Values* 25(4): 427–42.

Held, Virginia. 2006. *The Ethics of Care*. Oxford: Oxford University Press.

Helford, Reid. 1999. "Rediscovering the Presettlement Landscape: Making the Oak Savanna Ecosystem 'Real.'" *Science, Technology & Human Values* 24(1): 55–79.

Heneghan, Liam, Christopher Mulvaney, Kristen Ross, Lauren Umek, Cristy Watkins, Lynne Westphal, and David Wise. 2012. "Lessons Learned from Chicago Wilderness–Implementing and Sustaining Conservation Management in an Urban Setting." *Diversity* 4: 74–93.

Heneghan, Liam, Christopher Mulvaney, Kristen Ross, Susan Stewart, Lauren Umek, Cristy Watkins, Alaka Wali, Lynne Westphal, and David Wise. 2013. "Local Assessment of Chicago: From Wild Chicago to Chicago Wilderness." In *Urbanization, Biodiversity, and Ecosystem Services: Challenges and Opportunities*, 337–54. Dordrecht: Springer.

Heneghan, Liam, Lauren Umerk, Brad Bernau, Kevin Grady, Jamie Iatropulos, David Jabon, and Margaret Workman. 2009. "Ecological Research Can Augment Restoration Practice in Urban Areas Degraded by Invasive Species." *Urban Ecology* 12(1): 63–77.

Heneghan, Liam, Lynne Westphal, Kristen Ross, Cristy Watkins, Paul Gobster, Basil Iannone, Madeleine Tudor, Joanne Vining, Alaka Wali, Moira Zellner, and David Wise. 2019. "Institutional Diversity in the Planning Process Yields Similar Outcomes for Vegetation in Ecological Restoration." *Society & Natural Resources* 33(8): 949–67.

Herbert, Cassie. 2017. *Exclusionary Speech and Constructions of Community*. Dissertation, Georgetown University.

Hershenov, David. 1999. "Restitution and Revenge." *The Journal of Philosophy* 96(2): 79–94.

Hertog, Iris, and Ester Turnhout. 2018. "Ideals and Pragmatism in the Justification of Ecological Restoration." *Restoration Ecology* 26(6): 1221–9.

Hettinger, Ned. 2012. "Nature Restoration as a Paradigm for the Human Relationship with Nature." In *Ethical Adaptation to Climate Change*, edited by Allen Thompson and Jeremy Bendik-Keymer. Cambridge: MIT Press.

Heyward, Clare, and Dominic Roser. 2016. "Introduction." In *Climate Justice in a Non-Ideal World*, edited by Clare Heyward and Dominic Roser. Oxford: Oxford University Press.

Higgins, Leah. 2019. "A New Urbanism: Planning Cities at the Animal Scale." *Conference on Philosophy of the City*, Detroit, MI.

Higgs, Eric. 2003. *Nature by Design*. Cambridge: MIT Press.

Higgs, Eric. 2011. "Foreword." In *Human Dimensions of Ecological Restoration*, edited by Dave Egan, Evan Hjerpe, and Jesse Abrams. Washington, DC: Island Press.

Hilderbrand, Robert, Adam Watts, and April Randle. 2005. "The Myths of Restoration Ecology." *Ecology and Society* 10(1).

Hiller, Avram. 2011. "Climate Change and Individual Responsibility." *The Monist* 94: 349–68.

Hiskes, R. 2009. *The Human Right to a Green Future*. Cambridge: Cambridge University Press.

Hobbs, Richard. 2005. "The Future of Restoration Ecology." *Restoration Ecology* 13: 239–41.

Holifield, Ryan. 2001. "Defining Environmental Justice and Environmental Racism." *Urban Geography* 22: 78–90.
Holmgren, Margaret. 1983. "Punishment as Restitution." *Criminal Justice Ethics* 2: 36–49.
Holmgren, Margaret. 2012. *Forgiveness and Retribution*. Cambridge: Cambridge University Press.
Holtgren, Marty, Stephanie Ogren, and Kyle Whyte. 2014. "Renewing Relatives: Nmé Stewardship in a Shared Watershed." In *Tales of Hope and Caution in Environmental Justice*. Mellon Humanities for the Environment.
Holtgren, Marty, Stephanie Ogren, and Kyle Whyte. 2015. "Renewing Relatives: One Tribe's Efforts to Bring Back an Ancient Fish." *Earth Island Journal* 30(3): 54.
Hookway, Christopher. 2010. "Some Varieties of Epistemic Injustice." *Episteme* 7(2): 151–63.
Horsburgh, H. J. N. 1960. "The Ethics of Trust." *Philosophical Quarterly* 10: 343–54.
Hourdequin, Marion. 2012. "Restoration and History in a Changing World." *Ethics and the Environment* 18(2): 115–34.
Hourdequin, Marion. 2016. "Ecological Restoration, Continuity, and Change." In *Restoring Layered Landscapes*, edited by Marion Hourdequin and David Havlick. Oxford: Oxford University Press.
Hourdequin, Marion, and David Havlick. 2013. "Restoration and Authenticity Revisited." *Environmental Ethics* 35: 79–93.
Hourdequin, Marion, and David Wong. 2005. "A Relational Approach to Environmental Ethics." *Journal of Chinese Philosophy* 32: 19–33.
Hungate, Adam. 2005. *Let Them Eat Yellowcake*. Dissertation, University of California-Riverside.
Hyerczyk, Richard. 1993. "Comments: The Savanna Debate Rages." *Restoration & Management Notes* 11(2): 91.
Intergovernmental Panel on Climate Change (IPCC). 2013. *Climate Change 2013: Fifth Assessment Report*. Cambridge: Cambridge University Press.
James, William. 1896. *The Will to Believe and Other Essays in Philosophy*. New York: Dover.
Jamieson, Dale. 2002. *Morality's Progress*. Oxford: Oxford University Press.
Jamieson, Dale. 2014. *Reason in a Dark Time*. Oxford: Oxford University Press.
Johnson, Christopher. 2019. "Bringing Nature Back to the Chicago Area." *Earth Island Journal*, July 15.
Johnstone, Gerry. 2002. *Restorative Justice: Ideas, Values, Debates*. Devon: Willan Publishing.
Jones, Pattrice. 2010. "Roosters, Hawks, and Dawgs." *Feminism & Psychology* 20(3): 365–80.
Joravsky, Ben. 1996. "Up a Tree." *Chicago Reader*, November 7.
Jordan, William. 1986. "Restoration and the Reentry into Nature." *Restoration & Management Notes* 4: 2.
Jordan, William. 1987. "'Something We Were Always Looking For': The Origins of the Chicago Prairie Movement." *Restoration & Management Notes* 5: 68.

Jordan, William. 1991. "Shaping the Land, Transforming the Human Spirit." In *Helping Nature Heal*, edited by Richard Nilsen and Barry Lopez. Berkeley: Ten Speed Press.
Jordan, William. 1992. "Standing with Nature." *Restoration & Management Notes* 10: 111–12.
Jordan, William. 1995. "'Restoration' (The Word)." *Restoration & Management Notes* 13: 151.
Jordan, William. 1998. "Wilderness and Community." *Restoration & Management Notes* 16: 121.
Jordan, William. 2000. "Restoration, Community, and Wilderness." In *Restoring Nature*, edited by Paul Gobster and R. Bruce Hull. Washington, DC: Island Press.
Jordan, William. 2001. "Citizen – Or Member?" *Ecological Restoration* 19(1): 1–2.
Jordan, William. 2003. *The Sunflower Forest*. Berkeley: University of California Press.
Jordan, William. 2006. "Ecological Restoration." *Nature and Culture* 1(1): 22–35.
Jordan, William, and George Lubick. 2011. *Making Nature Whole: A History of Ecological Restoration*. Washington, DC: Island Press.
Jung, Courtney. 2018. "Reconciliation." *Journal of Global Ethics* 14(2): 252–65.
Katz, Eric. 1991. "The Ethical Significance of Human Intervention in Nature." *Restoration & Management Notes* 9: 90–6.
Katz, Eric. 1992. "The Big Lie." *Research in Philosophy and Technology* 12: 231–41.
Katz, Eric. 1993. "Response: The Savanna Debate Rages." *Restoration & Management Notes* 11.
Katz, Eric. 1996. "The Problem of Ecological Restoration." *Environmental Ethics* 18(2): 222–4.
Katz, Eric. 1997. *Nature as Subject*. Lanham: Rowman & Littlefield.
Katz, Eric. 2000. "Another Look at Restoration." In *Restoring Nature*, edited by Paul Gobster and R. Bruce Hull. Washington, DC: Island Press.
Katz, Eric. 2002a. "The Liberation of Humanity and Nature." *Environmental Values* 11: 397–405.
Katz, Eric. 2002b. "Understanding Moral Limits in the Duality of Artifacts and Nature: A Reply to Critics." *Ethics and the Environment* 7: 138–46.
Katz, Eric. 2012. "Further Adventures in the Case Against Restoration." *Environmental Ethics* 34(1): 67–97.
Katz, Eric. 2018. "Replacement and Irreversibility." *Ethics and the Environment* 23(1): 17–28.
Kendall, Peter. 1996. "Trouble in Prairieland." *Chicago Tribune*, October 6.
Kidd, Ian James, and Havi Carel. 2017. "Epistemic Injustice and Illness." *Journal of Applied Philosophy* 34(2): 172–90.
Kim, Claire Jean. 2015. *Dangerous Crossings*. Cambridge: Cambridge University Press.
Kimmerer, Robin Wall. 2002. "Weaving Traditional Ecological Knowledge into Biological Education." *BioScience* 52(5): 432–38.
Kimmerer, Robin Wall. 2011. "Restoration and Reciprocity." In *Human Dimensions of Ecological Restoration,* edited by Dave Egan, Evan Hjerpe, and Jesse Abrams. Washington, DC: Island Press.

Kimmerer, Robin Wall. 2013. *Braiding Sweetgrass*. Minneapolis: Milkweed.
Kincaid, Harold, John Dupre, and Alison Wylie, eds. 2007. *Value-Free Science? Ideals and Illusions*. Oxford: Oxford University Press.
King, Martin Luther. 1964. *Why We Can't Wait*. New York: Harper & Row.
Kitcher, Philip. 1993. *The Advancement of Science*. Oxford: Oxford University Press.
Klimchuk, Dennis. 2001. "Retribution, Restitution, and Revenge." *Law and Philosophy* 20: 81–101.
Korsgaard, Christine. 2018. *Fellow Creatures: Our Obligations to Other Animals*. Oxford: Oxford University Press.
Kukla, Rebecca. 2014. "Performative Force, Convention, and Discursive Justice." *Hypatia* 29(2): 440–57.
Lackey, Jennifer. 2016. "Pitted Against Yourself." *Blog of the APA*, April.
Ladkin, Donna. 2005. "Does 'Restoration' Necessarily Imply the Domination of Nature?" *Environmental Values* 14: 203–19.
LaDuke, Winona. 1994. "Traditional Ecological Knowledge and Environmental Futures." *Colorado Journal of International Environmental Law and Policy* 5(1): 127–48.
LaDuke, Winona. 1999. *All Our Relations*. Cambridge: South End Press.
LaDuke, Winona. 2005. *Recovering the Sacred*. Cambridge: South End Press.
Lake-Cook Audubon Chapter. 2020. "Montrose Point Bird Sanctuary." *Illinois Audubon Society*. https://www.lakecookaudubon.org/birding-sites/montrose-point-bird-sanctuary.
Lazarus, Richard. 2000. "'Environmental Racism! That's What It Is.'" *University of Illinois Law Review*: 255–74.
Le, Phuong. 2019. "Government's Controversial Owl Killings Spur Moral Questions About Human Intervention." *Associated Press*, October 19.
Lee, Alexander, Adam Perou Hermans, and Benjamin Hale. 2014. "Restoration, Obligation, and the Baseline Problem." *Environmental Ethics* 36(2): 171–86.
Lee, Charles. 1996. "Environment: Where We Live, Work, Play and Learn." *Race, Poverty, and the Environment* 6(2/3): 6–7.
Leopold, Aldo. 1966. *A Sand County Almanac, with Essays on Conservation from Round River*. New York: Ballantine Books.
Liddick, Donald R. 2006. *Eco-Terrorism: Radical Environmental and Animal Liberation Movements*. Westport: Praeger.
Light, Andrew. 1994. "Hegemony and Democracy." How Politics *in* Restoration Informs the Politics *of* Restoration." *Restoration & Management Notes* 12: 140–4.
Light, Andrew. 1995. "Urban Wilderness." In *Wild Ideas*, edited by David Rothenberg, 195–211. Minneapolis: University of Minnesota Press.
Light, Andrew. 2000a. "Ecological Restoration and the Culture of Nature." In *Restoring Nature*, edited by Paul Gobster and R. Bruce Hull. Washington, DC: Island Press.
Light, Andrew. 2000b. "Restoration, the Value of Participation, and the Risks of Professionalization." In *Restoring Nature*, edited by Paul Gobster and R. Bruce Hull. Washington, DC: Island Press.

Light, Andrew. 2003. "'Faking Nature' Revisited." In *The Beauty Around Us*, edited by Diane Michelfelder and William Wilcox. Albany: SUNY Press.
Light, Andrew. 2006. "Ecological Citizenship: The Democratic Promise of Restoration." In *The Human Metropolis*, edited by Rutherford Platt. Boston: University of Massachusetts Press.
Light, Andrew. 2008. "Restorative Relationships." In *Healing Natures, Repairing Relationships*, edited by Robert L. France. Sheffield: Green Frigate Press.
Light, Andrew. 2012. "The Death of Restoration?" In *Ethical Adaptation to Climate Change*, edited by Allen Thompson and Jeremy Bendik-Keymer. Cambridge: MIT Press.
Linzey, Andrew. 2009. *Why Animal Suffering Matters*. Oxford: Oxford University Press.
Linzey, Andrew, and Clair Linzey. 2018. *The Ethical Case Against Animal Experiments*. Urbana: University of Illinois Press.
Livezey, Kent B. 2010. "Killing Barred Owls to Help Spotted Owls I: A Global Perspective." *Northwestern Naturalist* 91(2): 107–33.
Lofton, Bonnie Price. 2004. "Transformative Justice." In *Critical Issues in Restorative Justice*, edited by Howard Zehr and Barb Toews. Cullompton: Willan Publishing.
Lomborg, Bjorn. 2007. *Cool It*. New York: Knopf.
MacClellan, Joel. 2013. "What the Wild Things Are." *Between the Species* 16(1): 6.
Machlis, Gary E. 1990. "The Tension Between Local and National Conservation Groups in the Democratic Regime." *Policy Review* 3: 278.
MacLachlan, Alice. 2008. "Forgiveness and Moral Solidarity." In *Forgiveness*, edited by Stephen Bloch-Shulman and David White, 3–16. Oxford: Inter-Disciplinary Press.
Marino, Lori, Gay Bradshaw, and Randy Malamud. 2009. "The Captivity Industry: The Reality of Zoos and Aquariums." *Best Friends Animal Society Magazine*, March–April.
Marsh, Gerald. 2011. "Testimony, Trust, and Prejudice in the Credibility Economy." *Hypatia* 26: 280–93.
Mason, Jim, and Peter Singer. 1990. *Animal Factories*. New York: Harmony Books.
Mason, Rebecca. 2011. "Two Kinds of Unknowing." *Hypatia* 26(2): 294–307.
Matheny, Gaverick. 2006. "Utilitarianism and Animals." In *In Defense of Animals: The Second Wave*, edited by Peter Singer, 13–25. Malden: Blackwell.
Matthews, Freya. 1999. "Letting the World Grow Old." *Worldviews: Global Religions, Culture, and Ecology* 3(2): 119–37.
Matthews, Freya. 2004. "Letting the World Do the Doing." *Australian Humanities Review* 33. Available online: http://australianhumanitiesreview.org/2004/08/01/letting-the-world-do-the-doing/.
McCormick, Hugh. 2009. "Intergenerational Justice and the Non-Reciprocity Problem." *Political Studies* 57(2): 451–8.
McGregor, Deborah. 2004. "Coming Full Circle: Indigenous Knowledge, Environment, and Our Future." *American Indian Quarterly* 28(3/4): 385–410.
McGregor, Deborah. 2008. "Linking Traditional Ecological Knowledge and Western Science." *Canadian Journal of Native Studies* 28(1): 139–58.

McGregor, Deborah. 2009. "Honoring Our Relations: An Anishnaabe Perspective." In *Speaking for Ourselves: Environmental Justice in Canada*, edited by Julian Agyeman, Peter Cole, and Randolph Haluza-Delay, 27–41. Vancouver: UBC Press.

McGregor, Deborah. 2018. "From 'Decolonized' to Reconciliation Research." *ACME: An International Journal for Critical Geographies* 17(3): 810–31.

McGurty, Eileen. 2007. *Transforming Environmentalism: Warren County, PCBs, and the Origins of Environmental Justice*. New Brunswick: Rutgers University Press.

McKenzie, Mia. 2014. "No More Allies." In *Black Girl Dangerous*. Oakland: BGD Press.

McKinnon, Rachel. 2016. "Epistemic Injustice." *Philosophy Compass* 11(8): 437–46.

McLaren, Duncan. 2018. "In a Broken World." *The Anthropocene Review* 5(2): 136–54.

McMahan, Jeff. 2008. "Eating Animals the Nice Way." *Daedalus* 137(1): 66–78.

McMahan, Jeff. 2010. "The Meat Eaters." *New York Times*, September 19.

Medina, Jose. 2011. "The Relevance of Credibility Excess in a Proportional View of Epistemic Injustice." *Social Epistemology* 25(1): 15–35.

Medina, Jose. 2013. *The Epistemology of Resistance*. Oxford: Oxford University Press.

Meekison, Lisa, and Eric Higgs. 1998. "The Rites of Spring (and Other Seasons): The Ritualizing of Restoration." *Restoration & Management Notes* 16: 73–81.

Mendelson, Jon. 1998. "Restoration from the Perspective of Recent Forest History." *Transactions of the Wisconsin Academy of Sciences, Arts, and Letters* 98: 137–48.

Mendelson, Jon, Stephen P. Aultz, and Judith Dolan Mendelson. 1992. "Carving Up the Woods: Savanna Restoration in Northeastern Illinois." *Restoration & Management Notes* 10: 127–31.

Mendelson, Jon, Stephen P. Aultz, and Judith Dolan Mendelson. 1993. "Comments: The Savanna Debate Rages." *Restoration & Management Notes* 11: 96–8.

Merchant, Carolyn. 2003. *Reinventing Eden*. New York: Routledge.

Meyer, Lukas. 2006. "Reparation and Symbolic Restitution." *Journal of Social Philosophy* 13(3): 406–22.

Meyer, Lukas. 2015. "Intergenerational Justice." In *Stanford Encyclopedia of Philosophy*, ed. Edward Zalta. Available online: https://plato.stanford.edu/entries/justice-intergenerational/.

Michaels, David. 2008. *Doubt Is Their Product*. Oxford: Oxford University Press.

Michaels, Patrick. 2011. *Climate Coup*. Washington, DC: The Cato Institute.

Miles, Irene, William Sullivan, and Frances Kuo. 1998. "Ecological Restoration Volunteers: The Benefits of Participation." *Urban Ecosystems* 2: 27–41.

Millard, Jere, Bruce Gallaher, David Baggett, and Steven Cary. 1983. "The Church Rock Uranium Mill Tailings Spill." *New Mexico Environmental Improvement Division*.

Mills, Charles. 2005. "'Ideal Theory' as Ideology." *Hypatia* 20(3): 165–84.

Mills, Stephanie. 1995. *In Service of the Wild*. Boston: Beason Press.

Minow, Martha. 1998. *Between Vengeance and Forgiveness*. Boston: Beacon Press.

Mintz-Woo, Kian. 2019. "Historical Responsibility for Loss & Damage." *International Society for Environmental Ethics Summer Conference*, Blue River, OR.

Mohai, Paul. 1994. "The Demographics of Dumping Revisited." *Virginia Environmental Law Journal* 14(4): 615–53.

Mohai, Paul. 1996. "Environmental Justice or Analytic Justice?" *Social Science Quarterly* 77(3): 500–7.

Mohai, Paul, David Pellow, and J. Timmons Roberts. 2009. "Environmental Justice." *Annual Review of Environment and Resources* 34: 405–30.

Moore, Kathleen Dean. 2016. *Great Tide Rising: Towards Clarity and Moral Courage in a Time of Planetary Change*. Berkeley: Counterpoint.

Moore, Michael. 1987. "The Moral Worth of Retribution." In *Responsibility, Character, and the Moral Emotions*, edited by Ferdinand Schoeman. Cambridge: Cambridge University Press.

Morris, Ruth. 2000. *Stories of Transformative Justice*. Toronto: Canadian Scholars' Press Inc.

Mouawad, Jad, and Christopher Jensen. 2015. "The Wrath of Volkswagen's Drivers." *New York Times*, September 21.

Murdock, Esme. 2016. *Ecological Reconciliation*. Dissertation, Michigan State University.

Murdock, Esme. 2018a. "Storied with Land: 'Transitional Justice' on Indigenous Lands." *Journal of Global Ethics* 14(2): 232–9.

Murdock, Esme. 2018b. "Unsettling Reconciliation." *Environmental Values* 27(5): 513–33.

Murdock, Esme. 2019. "Nature Where You're Not: Rethinking Environmental Spaces and Racism." In *The Routledge Handbook of Philosophy of the City*, edited by Sharon Meagher, Samantha Noll, and Joseph Biehl. New York: Routledge.

Murphy, Colleen. 2010. *A Moral Theory of Political Reconciliation*. Cambridge: Cambridge University Press.

Murphy, Jeffrie, and Jean Hampton. 1988. *Forgiveness and Mercy*. Cambridge: Cambridge University Press.

Nakashima, Douglas. 1993. "Astute Observers on the Sea Ice Edge." In *Traditional Ecological Knowledge*, edited by Julian Inglis. Ottawa: International Development Research Centre.

Nakashima, Douglas, Kirsty Galloway McLean, Hans Thulstrup, Ameyali Ramos Castillo, and Jennifer Rubis. 2012. *Weathering Uncertainty*. Paris: United Nations.

Nine, Cara. 2012. *Global Justice and Territory*. Oxford: Oxford University Press.

Nocella, Anthony J. 2011. "An Overview of the History and Theory of Transformative Justice." *Peace & Conflict Review* 6(1): 1–9.

Nolt, John. 2011. "How Harmful Are the Average American's Greenhouse Gases?" *Ethics, Policy, and Environment* 14(1): 3–10.

Norcross, Alistair. 2004. "Puppies, Pigs, and People." *Philosophical Perspectives* 18(4): 229–45.

Norgaard, Kari Marie. 2014a. "Social, Cultural, and Economic Impacts of Denied Access to Traditional Management." Report for the Karuk Tribe Department of Natural Resources.

Norgaard, Kari Marie. 2014b. "The Politics of Fire and the Social Impacts of Fire Exclusion on the Klamath." *Humboldt Journal of Social Relations* 36: 73–97.
Norlock, Kathryn J. 2009. *Forgiveness from a Feminist Perspective*. Lanham: Lexington Books.
Norlock, Kathryn J. 2010. "Forgiveness, Pessimism, and Environmental Citizenship." *Journal of Agricultural and Environmental Ethics* 23(1–2): 29–42.
Norlock, Kathryn J. 2017. "Ethical Hopelessness." *GenObs* 1(2): 29–39.
Norlock, Kathryn J. 2019. "Perpetual Struggle." *Hypatia* 34(1): 6–19.
North Branch Restoration Project. 2020. "About Us." https://northbranchrestoration.org/about-north-branch-restoration/.
O'Keefe, Joyce, Lenore Beyer-Chow, Jeremy Hojnicki, Holly Goldin, and Jennifer Welch. 2006. "Forest Preserve and Conservation District in Northeastern Illinois." *The Openlands Project*. https://www.csu.edu/cerc/documents/ForestPreserveDistrictStudy.pdf.
O'Keefe, William, and James Kueter. 2010. "Clouding the Truth: A Critique of *Merchants of Doubt*." *Marshall Institute Policy Outlook*, June. http://marshall.wpengine.com/wp-content/uploads/2011/06/OKeefe-and-Kueter-Clouding-the-Truth-A-Critique-of-Merchants-of-Doubt.pdf.
Oksanen, Markku. 2008. "Ecological Restoration as Moral Reparation." *Proceedings of the XXII World Congress of Philosophy* 23: 99–105.
Oreskes, Naomi, and Eric Conway. 2010a. *Merchants of Doubt*. New York: Bloomsbury.
Oreskes, Naomi, and Eric Conway. 2010b. "Defeating the Merchants of Doubt." *Nature* 465: 686.
Packard, Steve. 1985. "Rediscovering the Tallgrass Prairie." *Seventh Northern Illinois Prairie Workshop*, College of DuPage, June 1–2.
Packard, Steve. 1988. "Just a Few Oddball Species: Restoration and the Rediscovery of the Tallgrass Savanna." *Restoration & Management Notes* 6(1): 13–22. Reprinted in *Helping Nature Heal*, edited by Richard Nilsen and Barry Lopez. Berkeley: Ten Speed Press, 1991.
Packard, Steve. 1993. "Restoring Oak Ecosystems." *Restoration & Management Notes* 11: 5–16.
Packard, Steve. 1994. "Successional Restoration." *Restoration & Management Notes* 12: 32–9.
Packard, Steve. 2016. "After the Miracle." *Strategies for Stewards: From Woods to Prairies Blog*, May 26. https://woodsandprairie.blogspot.com/2016/05/after-miracle.html.
Packard, Steve. 2019. "Some History of Biodiversity Conservation." *Strategies for Stewards: From Woods to Prairies Blog*, April 26. http://woodsandprairie.blogspot.com/2019/04/some-history-of-biodiversity.html.
Page, Edward. 2006. *Climate Change, Justice, and Future Generations*. Cheltenham: Edward Elgar.
Palamar, Collette. 2006. "Restorashyn: Ecofeminist Restoration." *Environmental Ethics* 28(3): 285–301.
Palmer, Clare. 2010. *Animal Ethics in Context*. New York: Columbia University Press.

Palmer, Clare. 2011. "The Moral Relevance of the Distinction Between Domesticated and Wild Animals." In *Oxford Handbook on Animal Ethics*, edited by Tom Beauchamp and R. G. Frey, 701–25. Oxford: Oxford University Press.

Palmer, Clare. 2013. "What (If Anything) Do We Owe Wild Animals?" *Between the Species* 16(1): 4.

Palmer, Clare. 2018. "Should We Offer Assistance to Both Wild and Domesticated Animals?" *The Harvard Review of Philosophy* 25: 7–19.

Palmer, Clare. 2019. "Assisting Wild Animals Vulnerable to Climate Change." *Journal of Applied Philosophy* (Early View).

Palmer, Margaret, Joy Zedler, and Donald Falk. 2016. *Foundations of Restoration Ecology*, 2nd ed. Washington, DC: Island Press.

Palmer, Margaret, Richard Ambrose, and LeRoy Poff. 1997. "Ecological Theory and Community Restoration Ecology." *Restoration Ecology* 5(4): 291–300.

Pantazatos, Andreas. 2017. "Epistemic Injustice and Cultural Heritage." In *Routledge Handbook of Epistemic Injustice*, edited by Ian James Kidd, Gaile Pohlhaus Jr, and Jose Medina. London: Routledge.

Parrillo, Susan. 2008. "Fake Nature." *Proceedings of the XXII World Congress of Philosophy* 23: 123–30.

Pasternak, Judy. 2010. *Yellow Dirt: A Poisoned Land and the Betrayal of the Navajos.* New York: Free Press.

Pavlich, George. 2005. *Governing Paradoxes of Restorative Justice.* London: GlassHouse Press.

Peet, Andrew. 2015. "Epistemic Injustice in Utterance Interpretation." *Synthese* 194(9): 3421–43.

Pellow, David, and Robert J. Brulle, eds. 2005. *Power, Justice, and the Environment.* Cambridge: MIT Press.

Pepper, Angie. 2019. "Adapting to Climate Change." *Journal of Applied Philosophy* 36: 592–607.

Perry, Jonathan. 1994. "Greening Corporate Environments." *Restoration & Management Notes* 12: 145–57.

Petro, Debra Lynn. 1993. "Comments: The Savanna Debate Rages." *Restoration & Management Notes* 11(2): 94.

Phalen, Kimberly. 2009. "An Invitation for Public Participation in Ecological Restoration." *Ecological Restoration* 27(2): 178–86.

Philpott, Daniel. 2012. *Just and Unjust Peace.* Oxford: Oxford University Press.

Pierotti, Raymond, and Daniel Wildcat. 2000. "Traditional Ecological Knowledge: The Third Alternative." *Ecological Applications* 10(5): 1333–40.

Plumwood, Val. 1993. *Feminism and the Mastery of Nature.* New York: Routledge.

Pohlhaus Jr, Gaile. 2012. "Relational Knowing and Epistemic Injustice." *Hypatia* 27(4): 715–35.

Pohlhaus Jr, Gaile. 2017. "Varieties of Epistemic Injustice." In *Routledge Handbook of Epistemic Injustice*, edited by Ian James Kidd, Gaile Pohlhaus Jr, and Jose Medina. London: Routledge.

Posner, Eric, and David Weisbach. 2010. *Climate Change Justice.* Princeton: Princeton University Press.

Proctor, Robert. 2008. "Agnotology" In *Agnotology: The Making & Unmaking of Ignorance*, edited by Robert Proctor and Londa Schiebinger. Palo Alto: Stanford University Press.
Project R&R. 2019. "Sanctuaries." https://releasechimps.org/chimpanzees/sanctuary-facilities.
Putnam, Hilary. 2002. *The Collapse of the Fact-Value Dichotomy and Other Essays*. Cambridge: Harvard University Press.
Pyne, Andrew. 2017. *Fire in America*, 3rd ed. Seattle: University of Washington Press.
Radzik, Linda. 2004. "Making Amends." *American Philosophical Quarterly* 41(2): 141–54.
Radzik, Linda. 2009. *Making Amends: Atonement in Morality, Law, and Politics*. Oxford: Oxford University Press.
Raish, Carol. 2000. "Lessons for Restoration in the Tradition of Stewardship." In *Restoring Nature*, edited by Paul H. Gobster and R. Bruce Hall, 281–98. Washington, DC: Island Press.
Randall, Thomas. 2019. "Care Ethics and Obligations to Future Generations." *Hypatia* 34: 527–45.
Rassler, Steve. 1994. "Naturalness and Anthropocentricity." *Restoration & Management Notes* 12(2): 116–7.
Rawls, John. 1971. *A Theory of Justice*. Cambridge: Harvard University Press.
Rebik, Dana. 2020. "Neighbors near Legion Park Upset after Trees Cut Down for River Project." *WGN*, January 9.
Regan, Tom. 1982. *All That Dwell Therein: Animal Rights and Environmental Ethics*. Berkeley: University of California Press.
Regan, Tom. 1983. *The Case for Animal Rights*. Berkeley: University of California Press.
Regan, Tom. 1995. "Are Zoos Morally Defensible?" In *Ethics on the Ark*, edited by Bryan Norton, Michael Hutchins, Elizabeth F. Stevens, and Terry L. Maple. Washington, DC: Smithsonian.
Regan, Tom. 2001. *Defending Animal Rights*. Champaign: University of Illinois Press.
Regan, Tom. 2003. *Animal Rights, Human Wrongs*. Lanham: Rowman & Littlefield.
Regan, Tom. 2004a. *The Case for Animal Rights*, 2nd ed. Berkeley: University of California Press.
Regan, Tom. 2004b. *Empty Cages: Facing the Challenge of Animal Rights*. Lanham: Rowman & Littlefield.
Regan, Tom, and Gary Francione. 1992. "A Movement's Means Creates its Ends." *Animals' Agenda*, January/February.
Regan, Tom, and Peter Singer, eds. 1989. *Animal Rights and Human Obligations*. Cambridge: Cambridge University Press.
Robinson, Robert. 2018. "Global Environmental Justice." *Choice*, April 1–5.
Rodriguez, Karen, and Kurt Fuller. 2001. "Review of *Restoring Nature*, edited by Paul Gobster and R. Bruce Hall." *Ecological Restoration* 19(4): 221–4.

Rogers, Christina, and Eric Sylvers. 2015. "VW's Customers Feel Confusion, Remorse." *Wall Street Journal*, September 23.
Rohwer, Yasha, and Emma Marris. 2016. "Renaming Restoration." *Restoration Ecology* 24(5): 674–9.
Rollin, Bernard E. 1992. *Animal Rights and Human Morality.* Buffalo: Prometheus Books.
Ross, Laurel. 1992. "Thoughts on the Deer Problem." *Prairie Projections*, February 3–8.
Ross, Laurel. 1994. "Illinois' Volunteer Corps." *Restoration & Management Notes* 12(1): 5759.
Ross, Laurel. 1997. "The Chicago Wilderness." *Restoration & Management Notes* 15(1): 17–24.
Ross, Rupert. 2006. *Returning to the Teachings: Exploring Aboriginal Justice.* Toronto: Penguin.
Roy, Eleanor Ainge. 2017. "New Zealand River Granted Same Legal Rights as Human Being." *The Guardian*, March 16.
Ryder, Richard. 1975. *Victims of Science.* London: David Poynter.
Ryder, Richard. 2011. *Speciesism, Painism, and Happiness.* Exeter: Imprint Academic.
Sandweiss, Stephen. 1998. "The Social Construction of Environmental Justice." In *Environmental Injustices, Political Struggles*, edited by David E. Camacho. Durham: Duke University Press.
Sapontzis, Steve. 1987. *Morals, Reason, and Animals.* Philadelphia: Temple University Press.
Satz, Deborah. 2007. "Countering the Wrongs of the Past." In *Reparations*, edited by Jon Miller and Rahul Kumar, 176–92. Oxford: Oxford University Press.
Schaap, Andrew. 2005. *Political Reconciliation.* London: Routledge.
Scheffler, Samuel. 1997. "Relationships and Responsibilities." *Philosophy & Public Affairs* 26(3): 189–209.
Schlosberg, David. 2002. *Environmental Justice and the New Pluralism.* Oxford: Oxford University Press.
Schlosberg, David. 2004. "Reconceiving Environmental Justice." *Environmental Politics* 13(3): 517–40.
Schlosberg, David. 2007. *Defining Environmental Justice.* Oxford: Oxford University Press.
Schlosberg, David. 2013. "Theorising Environmental Justice." *Environmental Politics* 22: 37–55.
Schlosberg, David, and David Carruthers. 2010. "Indigenous Struggles, Environmental Justice, and Community Capabilities." *Global Environmental Politics* 10(4): 12–35.
Schott, Robin. 2004. "The Atrocity Paradigm and the Concept of Forgiveness." *Hypatia* 19(4): 202–9.
Schroeder, Herbert. 2000. "The Volunteer Experience." In *Restoring Nature*, edited by Paul H. Gobster and R. Bruce Hull. Washington, DC: Island Press.

Seavy, Nathaniel, Thomas Gardali, Gregory Golet, F. Thomas Griggs, Christine Howell, Rodd Kelsey, Stacy Small, Joshua Viers, and James Weigand. 2009. "Why Climate Change Makes Riparian Restoration More Important than Ever." *Ecological Restoration* 27(3): 330–8.

Sepinwall, Amy. 2006. "Responsibility for Historical Injustices." *Journal of Law and Politics* 22(3): 183–229.

Sessions, George. 1995. "Postmodernism and Environmental Justice." *Trumpeter*.

Sharp, Hasana. 2020. "Not All Humans." *Environmental Philosophy*, February 8 (Online First): 1–16.

Shephard, Arla, and Ray Ring. 2010. "'The Environment...Is Where We Live.'" *High Country News*, February 1.

Sher, George. 2005. "Transgenerational Compensation." *Philosophy & Public Affairs* 33: 181–200.

Shiva, Vandana. 1997. *Biopiracy*. Cambridge: South End Press.

Shiva, Vandana, and Radha Holla-Bhar. 1993. "Intellectual Piracy and the Neem Tree." *The Ecologist* 23(6): 223.

Shklar, Judith. 1990. *The Faces of Injustice*. New Haven: Yale University Press.

Shore, Debra. 1997. "Controversy Erupts Over Restoration in Chicago Area." *Restoration & Management Notes* 15(1): 25–31.

Shue, Henry. 1999. "Global Environment and International Inequality." *International Affairs* 75(3): 531–45.

Siceloff, Bruce. 2014. "Toxic Cleanup Shifts from Dirt near RDU to Region's Streams, Lakes." *The News & Observer*, July 19.

Siewers, Alf. 1998. "Making the Quantum-Cultural Leap: Reflections on the Chicago Controversy." *Restoration & Management Notes* 16(1): 9–15.

Siipi, Helena. 2003. "Artefacts and Living Artefacts." *Environmental Values* 12(4): 413–30.

Siipi, Helena. 2008. "Dimensions of Naturalness." *Ethics and the Environment* 13(1): 71–103.

Singer, Peter. 1974. "All Animals are Equal." *Philosophic Exchange* 5(1).

Singer, Peter. 1990. *Animal Liberation*, 2nd ed. London: Cape.

Singer, Peter. 2011. *Practical Ethics*, 3rd ed. Cambridge: Cambridge University Press.

Singer, S. Fred. 2000. "Cool Planet, Hot Politics." *American Outlook* (Summer): 38–40.

Sinnott-Armstrong, Walter. 2005. "It's Not My Fault: Global Warming and Individual Moral Obligations." In *Perspectives on Climate Change*, edited by Walter Sinnott-Armstrong and Ron Howarth. Amsterdam: Elsevier.

Smith, Anna. 2019. "The Klamath River Now Has the Legal Rights of a Person." *High Country News*, September 24.

Smith, Laura. 2014. "On the 'Emotionality' of Environmental Restoration: Narratives of Guilt, Restitution, Redemption, and Hope." *Ethics, Policy, and Environment* 17(3): 286–307.

Smith, Nick. 2008. *I Was Wrong*. Cambridge: Cambridge University Press.

Society for Ecological Restoration. 2020. "What Is Ecological Restoration?" https://www.ser-rrc.org/what-is-ecological-restoration/.

Spelman, Elizabeth. 2008. "Embracing and Resisting the Restorative Impulse." In *Healing Natures, Restoring Relationships*, edited by Robert L. France. Sheffield: Green Frigate Books.

Spinner-Halev, Jeff. 2012. *Enduring Injustice*. Cambridge: Cambridge University Press.

Steel, Daniel, and Kyle Whyte. 2012. "Environmental Justice, Values, and Scientific Expertise." *Kennedy Institute of Ethics Journal* 22(2): 163–82.

Stevens, Kathy. 2010. *Animal Camp*. New York: Skyhorse.

Stevens, William K. 1996. *Miracle Under the Oaks*. New York: Pocket Books.

Stoy, Andrew. 2015. "Why VW's Betrayal with Diesel Engines Is Different." *Automotive News*, September 21.

Strickland, Ruth Ann. 2004. *Restorative Justice*. New York: Peter Lang Publishing.

Strohmaier, David. 2000. "The Ethics of Prescribed Burns." *Ecological Restoration* 18(1): 5–9.

Sullivan, Dennis, and Larry Tifft. 2006. *Handbook of Restorative Justice*. London: Routledge.

Swink, Floyd. 1969. *Plants of the Chicago Region*. Morton Arboretum.

Swink, Floyd, and Gerould Wilhelm. 1994. *Plants of the Chicago Region*. Indianapolis, IN: Indiana Academy of Science.

Sze, Julie, and Jonathan K. London. 2008. "Environmental Justice at the Crossroads." *Sociology Compass* 2: 1331–54.

Tan, Kok-Chor. 2007. "Colonialism, Reparations, and Global Justice." In *Reparations*, edited by Jon Miller and Rahul Kumar, 280–306. Oxford: Oxford University Press.

Tanasescu, Mihnea. 2017. "Responsibility and the Ethics of Ecological Restoration." *Environmental Philosophy* 14(2): 255–74.

Taylor, Dorceta. 2000. "The Rise of the Environmental Justice Paradigm." *American Behavioral Scientist* 43(4): 508–80.

Taylor, Dorceta. 2014. *Toxic Communities: Environmental Racism, Industrial Pollution, and Residential Mobility*. New York: New York University Press.

Taylor, Paul. 1986. *Respect for Nature*. Princeton: Princeton University Press.

Taylor, Sunaura. 2017. *Beasts of Burden: Animal and Disability Liberation*. New York: New Press.

The Whale Sanctuary Project. 2019. "Our Work." https://whalesanctuaryproject.org/our-work.

Thomas, Jerry. 1991. "Other Species of Wildlife Pay Dear Price for Saving Deer." *Chicago Tribune*, October 15.

Thompson, Janna. 2001. "Historical Injustice and Reparation." *Ethics* 112(1): 114–35.

Thompson, Janna. 2002. *Taking Responsibility for the Past*. Cambridge: Polity Press.

Thompson, Janna. 2008. "Apology, Historical Obligations and the Ethics of Memory." *Memory Studies* 2(2): 195–210.

Thompson, Janna. 2009a. *Intergenerational Justice*. New York: Routledge.

Thompson, Janna. 2009b. "Identity and Obligation in a Transgenerational Polity." In *Intergenerational Justice*, edited by Axel Gosseries and Lukas Meyer, 25–49. Oxford: Oxford University Press.

Thompson, Janna. 2015. "Reparative Claims and Theories of Justice." In *Historical Justice and Memory*, edited by Klaus Newman and Janna Thompson. Madison: University of Wisconsin Press.

Throop, William. 1997. "The Rationale for Ecological Restoration." In *The Ecological Community*, edited by Roger Gottlieb. London: Routledge.

Throop, William. 2001. "Review of *Restoring Nature*." *Ecological Restoration* 19(4): 215–7.

Throop, William. 2012. "Environmental Virtues and the Aims of Restoration." In *Ethical Adaptation to Climate Change*, edited by Allen Thompson and Jeremy Bendik-Keymer. Cambridge: MIT Press.

Thunberg, Greta. 2019. "Speech at the UN Climate Action Summit." September 23. https://www.npr.org/2019/09/23/763452863/transcript-greta-thunbergs-speech-at-the-u-n-climate-action-summit.

Torpey, John. 2015. "The Political Field of Reparations." In *Historical Justice and Memory*, edited by Klaus Newman and Janna Thompson. Madison: University of Wisconsin Press.

Tsosie, Rebecca. 2012. "Indigenous Peoples and Epistemic Injustice." *Wash Law Review* 87: 1133.

Tutu, Desmond. 1999. *No Future Without Forgiveness*. New York: Doubleday.

Tuvel, Rebecca. 2015. "Sourcing Women's Ecological Knowledge." *Hypatia* 30(2): 319–36.

United States v. Ward. 1985. "618 Federal Supplement 884." September 9.

US Environmental Protection Agency (EPA). 1992. *Environmental Equity*. Washington, DC: US Environmental Protection Agency.

US Environmental Protection Agency (EPA). 2019. "Superfund Site: Ward Transformer Raleigh NC." https://cumulis.epa.gov/supercpad/cursites/csitinfo.cfm?id=0406082.

US Environmental Protection Agency (EPA). 2020. "Superfund Site: United Nuclear Corporation, Church Rock." https://cumulis.epa.gov/supercpad/cursites/csitinfo.cfm?id=0600819.

US Fish & Wildlife Service. 2013. "Final Decision Announced for Barred Owl Removal." https://www.fws.gov/pacific/news/news.cfm?id=2144375288.

US General Accounting Office. 1983. *Sitings of Hazardous Waste Landfills and their Correlation with Racial and Economic Status of Surrounding Communities*. Washington, DC: Government Print Office.

US House of Representatives. 1979. "Mill Tailings Dam Break at Church Rock, New Mexico." *Oversight Hearing Before the Subcommittee on Energy and the Environment*, October 22.

Van Matre, Lynn. 1996. "Forest Preserve Halts Tree Cutting." *Chicago Tribune*, October 2.

Vena, Natalie Bump. 2013. "Preservation's Loss: The Statutory Construction of Forests in Cook County." *Engagement Anthropology & Environment Society*

Blog. American Anthropological Association: https://aesengagement.wordpress.com/2013/12/17/preservations-loss-the-statutory-construction-of-forests-in-cook-county-il/.

Vena, Natalie Bump. 2014. "Early Prairie Restoration in the Forest Preserves of Cook County." Centennial History Series, Forest Preserves of Cook County. Originally posted June 3, 2014, updated May 22, 2019. https://fpdcc.com/centennial-history-series-early-prairie-restoration-in-the-forest-preserves-of-cook-county/.

Vena, Natalie Bump. 2016. *The Nature of Bureaucracy in the Cook County Forest Preserves.* PhD Dissertation, Department of Anthropology, Northwestern University.

Verdjea, Ernesto. 2009. *Unchopping a Tree.* Philadelphia: Temple University Press.

Vogel, Stephen. 2006. "The Silence of Nature." *Environmental Values* 15(2): 145–71.

Vogel, Stephen. 2015. *Thinking Like a Mall.* Cambridge: MIT Press.

Waldron, Jeremy. 1992. "Superseding Historic Injustice." *Ethics* 103(1): 4–28.

Walen, Alex. 2014. "Retributive Justice." In *Stanford Encyclopedia of Philosophy,* ed. Edward Zalta. Available online: https://plato.stanford.edu/entries/justice-retributive/.

Walker, Margaret Urban. 1998. *Moral Understandings.* New York: Routledge.

Walker, Margaret Urban. 2001. "Moral Repair and its Limits" In *Mapping the Ethical Turn,* edited by Todd Davis and Kenneth Womack. Charlottesville, VA: University of Virginia Press.

Walker, Margaret Urban. 2006a. *Moral Repair: Reconstructing Moral Relations After Wrongdoing.* Cambridge: Cambridge University Press.

Walker, Margaret Urban. 2006b. "Restorative Justice and Reparations." *Journal of Social Philosophy* 37(3): 377–95.

Walker, Margaret Urban. 2010. *What is Reparative Justice?* Milwaukee: Marquette University Press.

Walker, Margaret Urban. 2013. "The Expressive Burden of Reparations." In *Justice, Responsibility, and Reconciliation in the Wake of Conflict,* edited by Alice MacLachlan and Allen Speight. Dordecht: Springer.

Walker, Margaret Urban. 2014. "Moral Vulnerability and the Task of Reparations." In *Vulnerability,* edited by Catriona Mackenzie, Wendy Rogers, and Susan Dodds. Oxford: Oxford University Press.

Walker, Margaret Urban. 2015a. "How Can Truth Telling Count as Reparations?" In *Historical Justice and Memory,* edited by Klaus Newman and Janna Thompson. Madison: University of Wisconsin Press.

Walker, Margaret Urban. 2015b. "Making Reparations Possible: Theorizing Reparative Justice." In *Theorizing Transitional Justice,* edited by Claudio Corradetti, Nir Eisikovits, and Jack V. Rotondi. London: Ashgate.

Walker, Nigel. 1999. "Even More Varieties of Retribution." *Philosophy* 74(290): 595–605.

Wallace-Wells, David. 2019a. "It's Greta's World." *New York Magazine,* September 17.

Wallace-Wells, David. 2019b. *The Uninhabitable Earth*. New York: Tim Duggan.
Warren County v. North Carolina. 1981. "528 Federal Supplement 276." November 25.
Warren, Karen. 1990. "The Power and the Promise of Ecological Feminism." *Environmental Ethics* 12(2): 125–46.
Warren, Karen. 1999. "Environmental Justice: Some Ecofeminist Worries About a Distributive Model." *Environmental Ethics* 21(2): 151–61.
Wasserman, Harvey, Norman Solomon, Robert Alvarez, and Eleanor Walters. 1982. *Killing Our Own: The Disaster of America's Experience with Atomic Radiation*. New York: Random House.
Watkins, Cristy, Lynne Westphal, Paul Gobster, Joanne Vining, Alaka Wali, and Madeleine Tudor. 2015. "Shared Principles of Restoration Practice in the Chicago Wilderness." *Human Ecology Review* 21(1): 155–78.
Weiss, Edith Brown. 1992. "In Fairness to Future Generations and Sustainable Development." *American University International Law Review* 8(1): 19–26.
Wenar, Leif. 2006. "Reparations for the Future." *Journal of Social Philosophy* 37(3): 396–405.
Wenz, Peter. 1988. *Environmental Justice*. Albany: State University of New York Press.
Wenz, Peter. 2001. "Just Garbage." In *Faces of Environmental Racism: Confronting Issues of Global Justice*, edited by Laura Westra and Bill Lawson. Lanham: Rowman & Littlefield.
Westphal, Lynne, Cristy Watkins, Paul H. Gobster, Liam Heneghan, Kristen Ross, Laurel Ross, Madeleine Tudor, Alaka Wali, David H. Wise, Joanne Vining, and Moira Zellner. 2014. "Social Science Methods Used in the RESTORE Project." *General Technical Report NRS-138*, US Department of Agriculture, Forest Service, Northern Research Station: 1–116.
Wheeler, Samuel. 1997. "Reparations Reconstructed." *American Philosophical Quarterly* 34(3): 301–18.
Whyte, Kyle. 2011. "The Recognition Dimensions of Environmental Justice in Indian Country." *Environmental Justice* 4(4): 199–205.
Whyte, Kyle. 2013. "On the Role of Traditional Ecological Knowledge as a Collaborative Concept." *Ecological Processes* 2(7): 1–12.
Whyte, Kyle. 2016a. "Indigenous Experience, Environmental Justice, and Settler Colonialism." In *Nature and Experience*, edited by Bryan Bannon, 157–74. London: Rowman & Littlefield.
Whyte, Kyle. 2016b. "Indigeneity and US Settler Colonialism." In *Oxford Handbook of Philosophy and Race*, edited by Naomi Zack, 91–101. Oxford: Oxford University Press.
Whyte, Kyle. 2017a. "Is It Colonial Deja Vu?" *Humanities for the Environment*, edited by Joni Adamson, Michael Davis, and Hsinya Huang. London: Earthscan.
Whyte, Kyle. 2017b. "The Dakota Access Pipeline, Environmental Injustice, and US Colonialism." *Red Ink* 19(1): 154–69.
Whyte, Kyle. 2018a. "On Resilient Parasitisms, or Why I'm Skeptical of Indigenous/Settler Reconciliation." *Journal of Global Ethics* 14(2): 277–89.

Whyte, Kyle. 2018b. "What Do Indigenous Knowledges Do for Indigenous Peoples?" In *Traditional Ecological Knowledge*, edited by Melissa Nelson and Daniel Shilling. Cambridge: Cambridge University Press.

Whyte, Kyle. 2020. "Too Late for Indigenous Climate Justice." *WIRES Climate Change* 11(1): e603.

Wiegleb, Gerhard, Udo Boring, Gyewoon Choi, Hans-Uwe Dahms, Kamalaporn Kanongdate, Chan-Woo Byeon, and Lian Guey Ler. 2013. "Ecological Restoration as Precaution and Not as Restitutional Compensation." *Biodiversity and Conservation* 22(9): 1931–48.

Wilhelm, Gerould, and Laura Rericha. 2017. *Flora of the Chicago Region*. Indianapolis, IN: Indiana Academy of Science.

Wilk, Thomas. 2017. "Trust, Communities, and the Standing to Hold Accountable." *Kennedy Institute of Ethics Journal* 27(2): 1–22.

Williams, Florence. 1997. "Whither the Eco-warrior?" *Outside Magazine*, November.

Windhager, Steve. 1998. "Reply to Dwight Berry." *Restoration & Management Notes* 16(2): 128.

Woodworth, Paddy. 2013. *Our Once and Future Planet*. Chicago: University of Chicago Press.

Woolford, Andrew. 2009. *The Politics of Restorative Justice*. Halifax: Fernwood Publishing.

Wyckoff, Jason. 2014. "Linking Sexism and Speciesism." *Hypatia* 29(4): 721–37.

Young, Iris Marion. 1983. "Justice and Hazardous Waste." *The Applied Turn in Contemporary Philosophy: Bowling Green Studies in Applied Philosophy* 5: 171–83.

Yunger, John. 1992. "The Effects of Wetlands Restoration and Reconstruction on Mammalian Species Composition and Abundance." *Wetlands Research: Chicago* 6(4): 1–23.

Zehr, Howard. 1990. *Changing Lenses: A New Focus for Crime and Justice*. Scottdale: Herald.

Zehr, Howard. 2011. "Restorative or Transformative Justice?" *Restorative Justice Blog*, March 10. https://emu.edu/now/restorative-justice/2011/03/10/restorative-or-transformative-justice/.

Zehr, Howard. 2013. "Retributive Justice, Restorative Justice." In *A Restorative Justice Reader*, edited by Gerry Johnstone. New York: Routledge.

Zentner, John. 1992. "Commentary: The Issue of Restorability." *Restoration & Management Notes* 10(2): 113–5.

Zheutlin, Peter. 2015. *Rescue Road*. Naperville, IL: Sourcebooks.

Index

accountability, 8, 65, 83
acknowledgement of wrongdoing, 8, 41–43, 102–3
amelioration, 3–4, 8, 24
amends, 8, 13–14, 29, 42–43, 64–65, 103
American Indian Center, xn1, 127
American Indian knowledge. *See* indigenous knowledge
animal relationships. *See* interspecies relationships
animal rights, 47, 53–57
animal sanctuaries, 59–60
Anishinaabe, 34, 39, 66
anthropocentrism, 30–31, 47, 62
apology, 8, 29, 64–69
asynchronous. *See* intergenerational

Baier, Annette, 10, 58, 73–75, 84
Betz, Bob, 112–14, 132n8
biodiversity, 110, 122–23
biotic community, 10, 59, 138
Blix, Petra, 117–18, 122, 126
Bullard, Robert, 23

Caney, Simon, 71, 75
care ethics, 10, 58
Chavis, Benjamin, 22–23, 25
Chicago, Illinois, ix, 23, 110–11

Chicago Wilderness, 38, 109–32
Church Rock, New Mexico, 1, 87–91, 95
climate change, 69–84; compensation and reparations, 75–78; and intergenerational justice, 71–78; mitigation and adaptation, 76; refugees, 76; skepticism, 84n2
Coffey, Ray, 117–19, 121, 126–27, 133n23
community: ambiguities of, 123–24; cross-generational, 73–75
compensation: for climate change, 75–78; and environmental justice, 29; and epistemic justice, 103; and restitutive justice, 6–7, 59–60
contributory injustice, 94, 106
corrective justice, 4, 16n6, 72
credibility, 93–96

distributive justice, 4, 21, 28
Dotson, Kristie, 93–94, 105

ecofeminism, 58–59
ecological restoration, 37–50, 112–32; benevolent *vs.* malicious, 43; and community, 109, 123–24, 130–32; criticisms of, 41–43, 112–13; definitions of, 122, 134n28; and

171

faking nature, 41; historical authenticity, 40, 120–22; mitigation, 43; moratorium, 113, 119; and restoration ecology, 37, 122, 129; as ritual, 45, 109
Emmerman, Karen, 7, 9, 54, 59–60, 66, 132
environmental governance, 92–93, 99–100
environmental holism, 47–48
environmental justice, 20–35
Environmental Protection Agency (EPA), 1, 23, 26, 91, 99
environmental racism, 20–22
EPA. *See* Environmental Protection Agency
epistemic exploitation, 94, 106
epistemic governance injustice, 100–101
epistemic injustice, 88–105; amelioration of, 101–5; and traditional ecological knowledge, 94–101; varieties of, 93–94, 108n15
epistemic objectification, 93, 99
epistemic repair, 101–5
ethics of belief, 91, 107n4
ethics of care. *See* care ethics

feminism, 10, 58
fire, 111, 114, 127, 133n10
Forest Preserves of Cook County (FPCC), 111–26; restoration moratorium, 113, 119
forgiveness: epistemic, 104–5; intergenerational, 13, 80–81; interspecies, 12, 63–64; third party, 80, 85n12
FPCC. *See* Forest Preserves of Cook County
Fricker, Miranda, 91, 93–94

Gardiner, Steve, 73, 77
global warming. *See* climate change
greenhouse gas emissions, 69–70, 77
Greenpeace, 21, 23, 34

Gruen, Lori, 9, 17n25, 56, 58

Heneghan, Liam, 110, 121–22, 128–29
Hourdequin, Marion, 9, 40, 109

ideal *vs.* non-ideal theory, vii, 27, 50, 71–72, 101
ignorance: active *vs.* passive, 89, 107n2; hermeneutical, 94, 98–99
indigenous justice, 27, 33
indigenous knowledge, 88, 91–101, 128. *See also* traditional ecological knowledge
injustice: argumentative, 93, 95; contributory, 94, 106; discursive, 93, 98; epistemic, 88–105; governance, 100, 108n13; hermeneutical, 94, 98; interpretive, 93, 98; participatory, 93, 96–98; testimonial, 93–95, 105
intergenerational: forgiveness, 13, 80–81; justice, 69–84; relationships, 72–75, 80–84; trust, 12, 78, 80
Intergovernmental Panel on Climate Change (IPCC), 71
interspecies relationships, 58–64
IPCC. *See* Intergovernmental Panel on Climate Change

Jordan, William, 37, 45, 109, 123
justice: climate, 69–84; corrective, 4, 16n6, 72; distributive, 4, 21, 28; environmental, 20–35; epistemic, 101–5; indigenous, 27, 33; intergenerational, 69–84; participatory, 28, 131; recognition, 31, 131; reparative, 7–8, 61–62, 75–78; restitutive, 6–8, 41, 59–60, 77; restorative, 7–9, 27–30; retributive, 5–6, 25; transformative, 29–30

Katz, Eric, 37, 42, 48–49
Kimmerer, Robin, 9, 48–49, 95–98, 127, 132, 141–42
Korsgaard, Christine, 65, 139–40

LaDuke, Winona, 33–34
Leopold, Aldo, 10, 32, 137–38
Light, Andrew, 37, 43–44
locally undesirable land uses (LULU), 25

McGregor, Deborah, 31, 107n6
Mendelson, Jon, 115, 127, 133n10
Miracle Under the Oaks, 38, 113, 118–19
mitigation, 43, 76
moral repair, viii, 7, 17n22, 48–49
moral residues, 54, 60, 132
Murdock, Esme, 17n26, 22–23, 139

Navajo Nation, 1, 88–91, 95
Norlock, Kathryn, 84, 85n12
North Branch Restoration Project, 113–14, 118–21
Nuclear Regulatory Commission (NRC), 90, 95

Packard, Steve, 112–23, 133n24
Palmer, Clare, 60–62
Paris Agreement, 72
participatory justice, 28, 131
PIP. *See* the pure intergenerational problem
planetary rights, 72
pluralism, 28, 47–48
Potawatomi Nation, ix, 114, 120, 127
Principles of Environmental Justice, 23, 30
professionalization: in animal care, 14, 59–65; in ecological restoration, 14, 43–44
the pure intergenerational problem (PIP), 73, 77

Quinn, Mary Lou, 113, 117–18, 121, 126

Radzik, Linda, 10, 16n7, 17n24
recognition justice, 31, 131

reconciliation, 9, 17n22, 31–33, 139
Regan, Tom, 54–57
relational repair, viii, 7, 17n22, 48–49; ecological, 39–50; epistemic, 101–7; intergenerational, 72–84; interspecies, 58–66; wrongful, 14–15, 82–84
remediation, 5–6, 44–45
reparations, 7–8, 16nn17–18; and climate change, 75–78; and environmental justice, 26
reparative justice, 7–8, 61–62, 75–78
restitutive justice, 6–8, 41, 59–60, 76–77
restoration ecology, 37, 122, 129
restorative justice, 7–9, 27–30
retributive justice, 5–6, 25
Ross, Laurel, 111–12, 116–17

Schulenberg, Ray, 112, 114
scientific ecological knowledge (SEK), 92, 97–99
second-order wrongdoing, 3, 66, 105, 125–26
settler colonialism, 32, 39, 97–99
Shore, Debra, 116, 119, 134n38
Sierra Club, 21, 23, 25, 34–35
Southwest Organizing Project (SWOP), 21, 23
Swink, Floyd, 112, 115

TEK. *See* traditional ecological knowledge
testimonial injustice, 93–95, 105
testimonial quieting and smothering, 93, 98, 103
Thompson, Janna, 8, 17n21, 73
Thunberg, Greta, 70–71, 80, 84n3
toxic waste, 19–20, 22–23
traditional ecological knowledge, 87–107, 127–28; in the Chicago Wilderness, 127–28; as a collaborative concept, 91–93; and epistemic injustice, 94–101;

in relation to science, 92, 97–98, 107n11
transformative justice, 29–30
tribal knowledge. *See* indigenous knowledge
tribal sovereignty, 33, 100
trust, 7, 10, 126, 132; epistemic, 93, 99–100, 102; intergenerational, 12, 78, 80; interspecies, 12, 63–64

United Church of Christ (UCC), 21–22

victim, 15n3, 16n11; identification, 11–12, 46–48, 78–79; subjectivity, 28–29, 46, 79, 104, 130–32
Volkswagen (VW), 69–70, 84

Walker, Margaret Urban, vii–viii, 7–8, 11, 14, 16n6, 17n21, 28–29, 40, 42–44, 54, 62–63, 80
Warren County, North Carolina, 20–22
Whyte, Kyle, 31–32, 91–93, 99–100, 107n11, 141
wrongful repair, 14–15, 82–84

About the Author

Ben Almassi is an associate professor of Philosophy at Governors State University, on the southern edge of the Chicago Wilderness, where he teaches courses in philosophy, political theory, environmental ethics, medical ethics, feminist theory, and practical reasoning. He is also an affiliated faculty member in GSU's Gender & Sexuality Studies, Interdisciplinary Studies, and Political & Social Justice Studies programs, and GSU's new Certificate in Restorative Justice. His recent publications include "Epistemic Injustice and Its Amelioration" in *Social Philosophy Today*, "Ecological Restoration as Practices of Moral Repair" in *Ethics and the Environment*, "Climate Change and the Need for Intergenerational Reparative Justice" in *Journal of Agricultural and Environmental Ethics*, and "Toxic Funding: Conflicts of Interest and Their Epistemological Significance" in *Journal of Applied Philosophy*. In 2019–2020 he was a visiting fellow at the DePaul Humanities Center at DePaul University, and before joining the faculty at Governors State University in 2013, he was an associate professor of Philosophy and Humanities at the College of Lake County. Ben lives in Chicago with his partner Negin and their daughter Zeydi.

www.ingramcontent.com/pod-product-compliance
Lightning Source LLC
Chambersburg PA
CBHW050907300426
44111CB00010B/1423